Lecture Notes in Mathematics

A collection of informal reports and seminars
Edited by A. Dold, Heidelberg and B. Eckmann, Zürich

215

Peter L. Antonelli
Dan Burghelea
Peter J. Kahn

Institute for Advanced Study, Princeton, NJ/USA 1970

The Concordance-Homotopy Groups of Geometric Automorphism Groups

Springer-Verlag
Berlin · Heidelberg · New York 1971

Peter L. Antonelli
University of Alberta, Edmonton, Alberta/Canada

Dan Burghelea
Institut de Mathématique de l'Académie Roumaine, Bucarest/Roumanie

Peter J. Kahn
Cornell University, Ithaca, NY/USA

AMS Subject Classifications (1970): 57F99

ISBN 3-540-05560-6 Springer-Verlag Berlin · Heidelberg · New York
ISBN 0-387-05560-6 Springer-Verlag New York · Heidelberg · Berlin

Offsetdruck: Julius Beltz, Hemsbach/Bergstr.

INTRODUCTION

Let $\mathcal{A}$ be one of the following categories of oriented manifolds and orientation-preserving maps:

$\mathcal{D}iff$: C^∞ manifolds and diffeomorphisms.

$\mathcal{PL}$: PL manifolds and PL isomorphisms.

$\mathcal{T}op$: topological manifolds and homeomorphisms.

$\mathcal{H}$: topological manifolds and homotopy equivalences.

The $\mathcal{A}$-automorphisms of a closed, connected manifold M in $\mathcal{A}$ form a semi-group $\mathcal{A}(M)$ under composition, a group when $\mathcal{A} \neq \mathcal{H}$. We endow $\mathcal{A}(M)$ with the uniform C^∞ topology when $\mathcal{A} = \mathcal{D}iff$ and with the compact-open topology when $\mathcal{A} = \mathcal{T}op$ or $\mathcal{H}$. When $\mathcal{A} = \mathcal{PL}$, we give $\mathcal{A}(M)$ a PL structure, in the sense of Hirsch and Mazur [12], or we give it a semi-simplicial structure. In all cases, then, it becomes meaningful to study the homotopy theory of $\mathcal{A}(M)$ and, in particular, the homotopy groups $\pi_i(\mathcal{A}(M)) = \pi_i(\mathcal{A}(M); id_M)$, $i \geq 1$.

Unfortunately, these homotopy groups are, in general, very hard to compute. For example, when $\mathcal{A} = \mathcal{D}iff$, $i > 1$, and $\dim M > 3$, we know of no complete computation. Therefore, we weaken some of their structure.

A homotopy class in $\pi_i(\mathcal{A}(M))$ may be represented by an $\mathcal{A}$-automorphism of $M \times \mathbb{R}^i$ satisfying two conditions:

(1) It preserves the $\mathbb{R}^i$ parameters.

(2) It coincides with the identity outside some compact set.

A similar description applies to homotopies. We shall drop condition (1) for maps and homotopies. The result is the $i^{\underline{th}}$ <u>concordance-homotopy group</u> $\pi_i(\mathcal{Q};M)$ and a forgetful homomorphism

$$\Phi : \pi_i(\mathcal{Q}(M)) \to \pi_i(\mathcal{Q};M).$$

In a sense intended to be more suggestive than precise, the homomorphism Φ marks a passage from analysis to geometry.

It is our intention in this paper to study the groups $\pi_i(\mathcal{Q};M)$ with a view toward reducing their computation to problems in algebra, homotopy theory, and the theory of surgery on compact manifolds. We state two general conclusions here: namely, when M is 1-connected and $i \geq 1$, $\dim M + i \geq 6$, the groups $\pi_i(\mathcal{Q};M)$

(i) are finitely-generated abelian groups, and

(ii) depend, up to extension, only on the homotopy type of M.

By (ii) we mean that there is an exact sequence of abelian groups

$$E_i(M) \to F_i(M) \to \pi_i(\mathcal{Q};M) \to G_i(M) \to H_i(M)$$

in which E_i, F_i, G_i and H_i depend only on the homotopy type of M.

Assertions (i) and (ii) follow from our Exactness Theorem (3.3.2), our Classification Theorem (3.5.5), and results in Appendix D. Assertion (i) remains valid when M is not 1-connected, provided that the Wall groups $L_k^s(\pi_1(M))$ are finitely-generated (e.g., when $\pi_1(M)$ is free, or abelian, or finite). Assertion (ii) holds with no conditions on $\pi_1(M)$.

Except in very special cases, nothing is known about the kernel of the forgetful homomorphism Φ. On the other hand, when $\mathcal{A} = \mathcal{D}iff$, the authors have demonstrated the non-triviality of image Φ for various kinds of M, [1]. This paper can be considered as a sequel to [1], although totally different in spirit, because we apply some of our general theory in order to determine non-trivial image Φ when $\mathcal{A} = \mathcal{D}iff$ (Appendix E). We use these results to deduce our announced Theorem C of [1] (see the Introduction of [1]). The numbering of the text and the appendices reflects this sequential concept: thus, this text consists of Chapter 3 and Appendices C-E; Chapters 1 and 2 and Appendices A and B appear in [1]. On the other hand, except for Appendix E, this paper is self-contained and entirely independent of [1].

We now give a brief summary.

First, we define the groups $\pi_i(\mathcal{A};M)$ (in a slightly more general form) and "relative" groups $\pi_{i+1}(\mathcal{B},\mathcal{A};M)$ (§§3.1, 3.2). Here, $(\mathcal{B},\mathcal{A})$ is any one of the pairs of categories $(\mathcal{T}op, \mathcal{D}iff)$, $(\mathcal{H}, \mathcal{D}iff)$, $(\mathcal{T}op, \mathcal{PL})$, $(\mathcal{H}, \mathcal{PL})$, $(\mathcal{H}, \mathcal{T}op)$. We then obtain an exact sequence of groups and homomorphisms (§3.3)

(1) $$\cdots \to \pi_{i+1}(\mathcal{B},\mathcal{A};M) \to \pi_i(\mathcal{A};M) \to \pi_i(\mathcal{B};M) \to \cdots$$

with suitable naturality properties. This sequence is also obtained by Hodgson [14], [15] who studies these homotopy groups from a somewhat different point of view.

Our key result is a classification theorem (§3.5), which, when $\mathcal{B} \neq \mathcal{H}$, can be expressed, in a slightly special case, as follows: if $\dim M + i \geq 6$, $i \geq 1$, there exists a natural isomorphism

$$(2) \qquad \pi_i(\mathcal{B}, \mathcal{A}; M) \approx [\Sigma^i(M \cup \text{point}), B/A]$$

where the pair (B, A) of "stable groups" is (TOP, O), (G, O), (TOP, PL), (G, PL), or (G, TOP), depending on the value of $(\mathcal{B}, \mathcal{A})$. When $\mathcal{B} = \mathcal{H}$, the statement is slightly more complicated and generalizes a result of D. Sullivan [33] (see also [15]).

When $(\mathcal{B}, \mathcal{A}) = (\mathcal{PL}, \mathcal{Diff})$, everything goes through, but the constructions and proofs require non-trivial results from Whitehead C^1-triangulation theory. For example, when $(\mathcal{B}, \mathcal{A})$ is one of the pairs listed above, every $\mathcal{A}$-map is <u>a fortiori</u> a $\mathcal{B}$-map. This immediately determines the homomorphism

$$(3) \qquad \pi_i(\mathcal{A}; M) \to \pi_i(\mathcal{B}; M)$$

of sequence (1). When $(\mathcal{B}, \mathcal{A}) = (\mathcal{PL}, \mathcal{Diff})$, however, the corresponding fact is not true. This kind of problem was first faced in [12] and [24], and we use their solution here. Namely, we use PD (piecewise-differentiable) maps instead of PL maps. Since every $\mathcal{Diff}$-map is PD, and since the space of PL maps is a "deformation retract" of the space of PD maps, we again obtain (3). Other less easily resolvable problems occur in connection with the relative groups $\pi_i(\mathcal{PL}, \mathcal{Diff}; M)$. Finally, we note that, given the $\mathcal{D}iff$-manifold M, we must choose a PL structure for M. According to Whitehead [39], we may choose a PL structure that contains a smooth triangulation, and any two such structures on M are PL isomorphic. Unfortunately, the isomorphism is not canonical. Thus, it is not a priori clear that, when $(\mathcal{B},\mathcal{Q}) = (\mathcal{PL}, \mathcal{D}iff)$, the sequence (1) is independent of the choice of smooth triangulation. All these problems are resolved in Appendix C, and the main program of the paper is carried out there for the $(\mathcal{PL}, \mathcal{D}iff)$ case.

Finally, in §4 and Appendix D, we compute a number of groups $\pi_i(\mathcal{Q};M)$ (see also the corollaries in §5).

TABLE OF CONTENTS

Peter L. Antonelli and Dan Burghelea were supported in part by National Science Foundation grant GP-7952X1 and Peter J. Kahn in part by a National Science Foundation Postdoctoral Fellowship.

Chapter 3

THE CONCORDANCE-HOMOTOPY GROUPS OF GEOMETRIC AUTOMORPHISM GROUPS

§3.1 PRELIMINARY DEFINITIONS AND LEMMAS

3.1.1 The Geometric Categories

All the objects in the following categories will be oriented manifolds, and we assume that all the morphisms preserve orientation:

$\mathcal{D}iff$: C^{∞} manifolds and diffeomorphisms.*

$\mathcal{PL}$: PL manifolds and PL isomorphisms.

$\mathcal{T}op$: Topological manifolds and homeomorphisms.

$\mathcal{H}$: Topological manifolds and homotopy equivalences.

Note that there are forgetful functors

$$\begin{array}{ccccc} & & \mathcal{PL} & & \\ & & \searrow & & \\ \mathcal{D}iff & \longrightarrow & \mathcal{T}op & \longrightarrow & \mathcal{H}. \end{array}$$

By a pair of geometric categories $(\mathcal{B}, \mathcal{A})$, we mean a pair of the above categories $\mathcal{A}$ and $\mathcal{B}$ for which there exists a forgetful functor $\mathcal{A} \longrightarrow \mathcal{B}$ as above. In this case, we may write $\mathcal{A} < \mathcal{B}$.

If M is an object of $\mathcal{A}$, and if $\mathcal{A} < \mathcal{B}$, then we shall consider M to be an object of $\mathcal{B}$ as well by remembering only its $\mathcal{B}$-structure.

3.1.2 Products and corners

The categories $\mathcal{PL}$, $\mathcal{T}op$, and $\mathcal{H}$ are clearly closed under the operation of cartesian product of manifolds and maps. We shall find it

---

* See 3.1.2

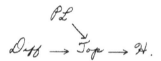

convenient to assume that $\mathcal{D}iff$ is also closed under this operation. Note that the natural C^∞ structure on a product of C^∞ manifolds will have "corners" whenever more than one factor has non-empty boundary. A general treatment of manifolds-with-corners is given in [6] or in [7]. For our purposes, however, it will suffice to take for the objects of $\mathcal{D}iff$ the smallest class of C^∞ manifolds-with-corners that contains the usual C^∞ manifolds and is closed with respect to cartesian product and the procedure of taking an open subset. We use the following definition of diffeomorphism:

Given C^∞ manifolds M_i, $i = 1, \ldots, k$ (resp., M'_j, $j = 1, \ldots, \ell$), choose a C^∞ imbedding $M_i \subset E_i$ (resp., $M'_j \subset E'_j$), where E_i (resp., E'_j) is a Euclidean space of dimension at least $2 \dim M_i + 2$ (resp., $2 \dim M'_j + 2$) and $\sum_i \dim E_i = \sum_j \dim E'_j$. Then, a C^∞ diffeomorphism

$$M_1 \times \ldots \times M_k \longrightarrow M'_1 \times \ldots \times M'_\ell$$

is a homeomorphism which extends to a diffeomorphism of a neighborhood of $M_1 \times \ldots \times M_k$ in $E_1 \times \ldots \times E_k$ onto a neighborhood of $M'_1 \times \ldots \times M'_\ell$ in $E'_1 \times \ldots \times E'_\ell$.

An analogous definition applies to open subsets of such cartesian products.

It is easy to verify that this definition has the following properties: (i) it is independent of the choices of imbeddings $M_i \subset E_i$ and $M'_j \subset E'_j$; (ii) compositions and inverses of diffeomorphisms are diffeomorphisms; (iii) when there are no corners, the definition coincides with the usual one; (iv) cartesian products of diffeomorphisms are again diffeomorphisms.

Thus, $\mathcal{D}iff$ consists of C^∞ manifolds and diffeomorphisms in the above sense.

We shall have occasion in later sections to refer to an "$\mathcal{A}$-submanifold" of an $\mathcal{A}$-manifold M. When $\mathcal{A} \neq \mathcal{D}iff$ this has the obvious meaning. When $\mathcal{A} = \mathcal{D}iff$ we use the following convention: the $\mathcal{D}iff$ submanifolds form the smallest class of topological submanifolds of objects in *Diff* containing the usual C^∞ submanifolds (of manifolds without corners) and closed under cartesian products, open subsets, and diffeomorphism. Thus, for example, if $N_i \subset M_i$ is a *Diff*-submanifold, in the above sense, and if there is a diffeomorphism

$$M_1 \times \ldots \times M_n \xrightarrow{f} M$$

in *Diff*, then $f(N_1 \times \ldots \times N_n) \subset M$ is a *Diff*-submanifold.

One problem with the above definitions and conventions is that the boundary of a manifold-with-corners fails to have a canonical C^∞ structure near the corners. There is, however, a well-known procedure for "straightening" corners in product manifolds (e.g., see [7] or [26]), which depends on a choice of C^∞ collars for the boundaries of the factors. The result is a C^∞ manifold-with-boundary whose diffeomorphism type is independent of the choice of collars. Thus, boundaries can be given (non-canonical) meaning in *Diff*. Unfortunately, the straightening procedure may introduce singularities in our diffeomorphisms (at corners). In particular, the product of diffeomorphisms will, in general, no longer be a diffeomorphism on the straightened product manifold. However, if the diffeomorphisms are suitably nicely behaved with respect to the distinguished collars, then they remain diffeomorphisms of the straightened product manifolds. Below, we give a criterion for such "nice behavior," and in Lemma 3.3.7, we show that, when needed, we may assume this criterion to be fulfilled without loss of generality.

3.1.3 Stationary maps

Let $f: (M, N) \longrightarrow (M, N)$ be a map of pairs and let $c: N \times [0, 1] \longrightarrow M$ be a map satisfying $c(x, 0) = x$, for all $x \in N$. Then, we say that f is stationary with respect to c if

$$f(c(x, t)) = c(f(x), t), \quad \text{for all } (x, t) \in N \times [0, 1].$$

For any ε, $0 < \varepsilon \leq 1$, let c_ε be given by $c_\varepsilon(x, t) = c(x, \varepsilon t)$. We say that f is <u>stationary near</u> N with respect to c if it is stationary with respect to c_ε, for some ε as above.

We omit reference to c when M is of the form $M_0 \times [0, 1]$, N is one of the ends of this cylinder, and c is given by the product structure.

Similar definitions apply to maps

$$f: (M, N) \longrightarrow (M', N')$$

and $c: N \times [0, 1] \longrightarrow M$, $c': N' \times [0, 1] \longrightarrow M'$, such that $c(x, 0) = x$, for all $x \in N$, and $c'(x', 0) = x'$, for all $x' \in N'$.

As an application of these notions, let the C^∞ manifolds M_i be given and let

$$M_1 \times \ldots \times M_n \xrightarrow{f} M_1 \times \ldots \times M_n$$

be a diffeomorphism. Choose a C^∞ collar

$$c_i: \partial M_i \times [0, 1] \longrightarrow M_i, \; i = 1, \ldots, n,$$

let N_i be obtained from the product $M_1 \times \ldots \times M_n$ by replacing M_i by ∂M_i, and define

$$\hat{c}_i: N_i \times [0, 1] \longrightarrow M_1 \times \ldots \times M_n$$

by

$$\hat{c}_i((x_1, \ldots, y_i, \ldots, x_n), t) = (x_1, \ldots, c_i(y_i, t), \ldots, x_n).$$

Now suppose that

a) $f(N_i) \subseteq N_i$, and

b) f is stationary near N_i with respect to $\hat{c}_i$, for all $i = 1, \ldots, n$.

Then, if we straighten the product $M_1 \times \ldots \times M_n$ via the collars $c_1, \ldots, c_n$, f becomes a self-diffeomorphism of the resulting straightened manifold.

3.1.4 The $(\mathcal{PL}, \mathcal{D}iff)$ case

Note that there is no forgetful functor

$$\mathcal{D}iff \longrightarrow \mathcal{PL}$$

so that the pair $(\mathcal{PL}, \mathcal{D}iff)$ is not a pair of geometric categories in the sense of 3.1.1. Nevertheless, all of our general results on pairs of geometric categories apply to $(\mathcal{PL}, \mathcal{D}iff)$ as well, and the $(\mathcal{PL}, \mathcal{D}iff)$ case will be useful for our applications later. Unfortunately, the proofs in this case involve a considerable amount of technical material from J. H. Whitehead's PD triangulation theory. Since this would distract from our main development, we present the $(\mathcal{PL}, \mathcal{D}iff)$ case separately in Appendix C. This appendix also contains some results in PD triangulation theory that may be of independent interest.

3.1.5 Supports

Let $f: Z \longrightarrow Z$ be a self-map of the topological space Z, and define $C \subseteq Z$ by

$$C = \{z \in Z \mid f(z) \neq z\}.$$

Then, the support of f, denoted by supp f, is the closure of $f^{-1}fC$. The following relations, which are not hard to verify, will be important:

$$f(\text{supp } f) = \text{supp } f = f^{-1}f(\text{supp } f).$$

3.1.6 The Automorphism Groups

Let $(\mathcal{B}, \mathcal{A})$ be a pair of geometric categories, $\mathcal{B} \neq \mathcal{H}$, let M be an object of $\mathcal{A}$, let X be a closed subset of M, and let N be a codimension-0, locally-flat, $\mathcal{A}$-submanifold of ∂M, closed as a subset of ∂M.* We define the automorphism group

$$\mathcal{A}(M \text{ rel } X)$$

to consist of all maps $M \longrightarrow M$ in $\mathcal{A}$ <u>with compact support</u> that avoids X. We define

$$(\mathcal{B}, \mathcal{A})(M, N \text{ rel } X)$$

to consist of all maps $f \in \mathcal{B}(M \text{ rel } X)$ for which $f(N) = N$ and $f|U\colon U \longrightarrow f(U)$ is a map in $\mathcal{A}$, for some neighborhood U of N in M (the neighborhood depending on f). Clearly, both $\mathcal{A}(M \text{ rel } X)$ and $(\mathcal{B}, \mathcal{A})(M, N \text{ rel } X)$ are groups with respect to map composition.

An $\mathcal{A}$-<u>concordance</u> rel X between maps f_0, f_1 in $\mathcal{A}(M \text{ rel } X)$ is a map

$$F \in \mathcal{A}(M \times [0, 1] \text{rel } X \times [0, 1])$$

such that

$$F(x, t) = (f_i(x), t), \quad i = 0, 1,$$

when $|t-i| < \varepsilon$, for some $\varepsilon > 0$. In this case, we say that F <u>is stationary within</u> ε <u>of</u> $\{0, 1\}$. A $(\mathcal{B}, \mathcal{A})$-<u>concordance</u> rel X is defined analogously. It is a map in $(\mathcal{B}, \mathcal{A})(M \times [0, 1], N \times [0, 1] \text{rel } X \times [0, 1])$ stationary within some $\varepsilon > 0$ of $\{0, 1\}$.

These notions of concordance clearly define equivalence relations on the corresponding automorphism groups. The relations are compatible with the group operations. The corresponding factor groups will be denoted by

---

* As mentioned in 3.1.2, when $\mathcal{A} = \mathcal{D}iff$ and M has corners, ∂M does not have a canonical C^∞ structure. Thus, to avoid complications in this case, we allow only those N of the form $\partial M \cap W$, where W is a codimension-0 $\mathcal{D}iff$-submanifold of M, in the sense of 3.1.2.

$$\pi_0(\mathcal{A};\ M \text{ rel } X) \text{ and}$$
$$\pi_0(\mathcal{B}, \mathcal{A};\ M, N \text{ rel } X).$$

3.1.7 <u>The semigroups</u> $\mathcal{H}$(M rel X) <u>and</u> ($\mathcal{H}$, $\mathcal{A}$)(M, N rel X)

This case must be treated separately and at greater length to take into account the fact that the morphisms in $\mathcal{H}$ are not, in general, isomorphisms.

First, corresponding to the pair of geometric categories ($\mathcal{H}$, $\mathcal{A}$), we define semigroups and notions of concordance rel X exactly as in 3.1.6, denoting them by

$$\mathcal{H}'(M \text{ rel } X),$$
$$(\mathcal{H}', \mathcal{A})(M, N \text{ rel } X),$$

$\mathcal{H}'$-concordance rel X, and ($\mathcal{H}'$, $\mathcal{A}$)-concordance rel X, respectively. As in 3.1.6, we obtain "factor" semigroups

$$\pi_0(\mathcal{H}';\ M \text{ rel } X)$$
$$\pi_0(\mathcal{H}', \mathcal{A};\ M, N \text{ rel } X).$$

Let

$$\mathcal{H}(M \text{ rel } X) \subseteq \mathcal{H}'(M \text{ rel } X) \text{ and}$$
$$(\mathcal{H}, \mathcal{A})(M, N \text{ rel } X) \subseteq (\mathcal{H}', \mathcal{A})(M, N \text{ rel } X)$$

be the sub-semigroups which consist of all pre-images of units in

$$\pi_0(\mathcal{H}';\ M \text{ rel } X)$$
$$\pi_0(\mathcal{H}', \mathcal{A};\ M, N \text{ rel } X),$$

respectively.

We then define the notions of $\mathcal{H}$-concordance rel X and ($\mathcal{H}$, $\mathcal{A}$)-concordance rel X on these sub-semigroups just as before. Such concordances will be maps in $\mathcal{H}(M \times [0, 1] \text{ rel } X \times [0, 1])$ and $(\mathcal{H}, \mathcal{A})(M \times [0, 1], N \times [0, 1] \text{ rel } X \times [0, 1])$, respectively, stationary near $\{0, 1\}$. The corresponding "factor" semigroups will be denoted by

$$\pi_0(\mathcal{H}; M \text{ rel } X) \text{ and}$$
$$\pi_0(\mathcal{H}, \mathcal{A}; M, N \text{ rel } X).$$

3.1.8 Lemma: $(\mathcal{H}', \mathcal{A})$-concordance rel X and $(\mathcal{H}, \mathcal{A})$-concordance rel X coincide on $(\mathcal{H}, \mathcal{A})(M, N \text{ rel } X)$.

More precisely, every $(\mathcal{H}', \mathcal{A})$-concordance rel X between maps in $(\mathcal{H}, \mathcal{A})(M, N \text{ rel } X)$ is an $(\mathcal{H}, \mathcal{A})$-concordance rel X.

We prove this at the end of this section. Observe that this implies that

$$\pi_0(\mathcal{H}; M \text{ rel } X) \text{ and}$$
$$\pi_0(\mathcal{H}, \mathcal{A}; M, N \text{ rel } X)$$

coincide with the sub-semigroups of units in

$$\pi_0(\mathcal{H}'; M \text{ rel } X) \text{ and}$$
$$\pi_0(\mathcal{H}', \mathcal{A}; M, N \text{ rel } X),$$

respectively. (The first case follows trivially from the second by specializing $N = \emptyset$.) In particular, they are groups.

We shall prove 3.1.8 with the aid of the following sublemma which will also have useful application in §3.3. Let $T: [0, 1] \times [0, 1] \longrightarrow [0, 1] \times [0, 1]$ be given by $T(r, s) = (s, r)$.

3.1.9 Sublemma: Let $(\mathcal{B}, \mathcal{A})$ be a pair of geometric categories (or $\mathcal{B}$ can denote $\mathcal{H}'$). Suppose that A and B are $(\mathcal{B}, \mathcal{A})$-concordances rel X, both stationary within $\frac{1}{3}$ of $\{0, 1\}$, such that $A_1 = B_0$. Let $S = [0, 1] \times [0, 1]$, and define $C: M \times \partial S \longrightarrow M \times \partial S$ by (cf. diagram below):

$$C(x, r, s) = \begin{cases} (A(x, r), 0), & s = 0 \\ (id_M \times T)(A(x, s), 0), & r = 0 \\ (B(x, r), 1), & s = 1 \\ (id_M \times T)(B(x, s), 1), & r = 1 \end{cases}$$

Then, C extends to a map in $(\mathcal{B}, \mathcal{A})(M \times S, N \times S \text{ rel } X \times S)$ which is stationary near $M \times \partial S$ with respect to the product structure.

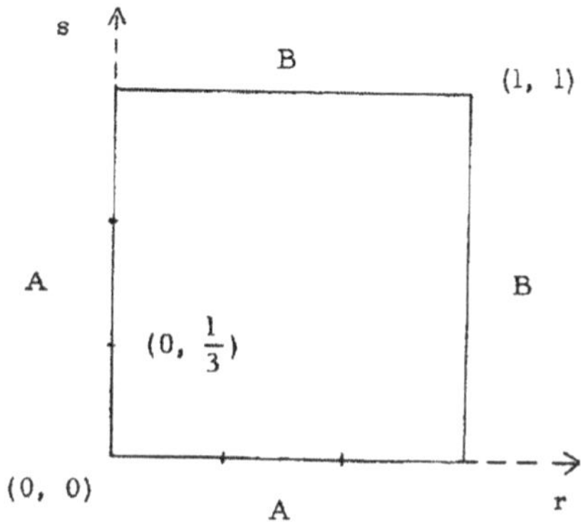

Proof: Let $S' = [\frac{1}{4}, \frac{3}{4}] \times [\frac{1}{4}, \frac{3}{4}]$ and extend C trivially to $M \times (S - \text{int } S')$.

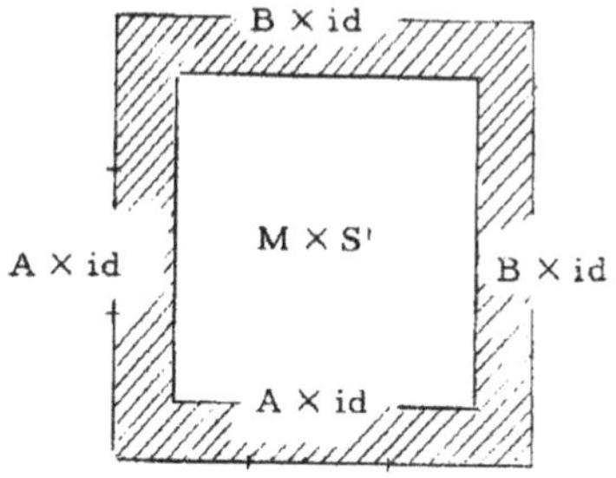

The postulated stationariness of A and B, together with the fact that $B_0 = A_1$, implies that the extension is well defined.

Assume for the moment that $\mathcal{A} = \mathcal{PL}$ or $\mathcal{T}op$. The proof consists of constructing an $\mathcal{A}$-automorphism $h : S' \longrightarrow S'$ so that $(id_M \times h)C(id_M \times h^{-1}) | M \times \partial S'$ has an obvious extension as a product map of the form $D \times id_{[0, 1]}$. We then fill in $M \times S'$ above by $(id_M \times h^{-1})(D \times id_{[0, 1]})(id_M \times h)$. This gives the desired extension of C.

We now construct h. Consider the following diagram.

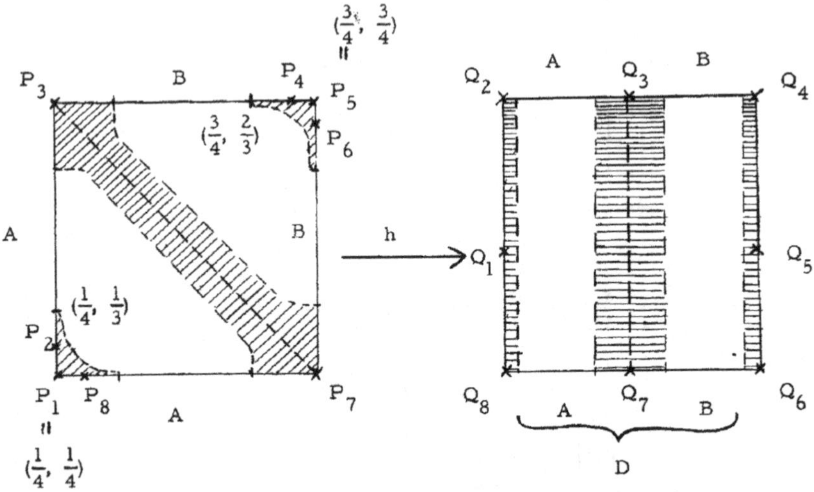

On $\partial S'$, h is defined by sending $P_i \longrightarrow Q_i$, $i = 1, \ldots, 8$, and extending linearly. Then extend h to S' by coning from the center. The shaded areas indicate schematically what happens near corners and near the major diagonal. Note that since A and B are stationary within $\frac{1}{3}$ of $\{0, 1\}$, the restriction of $(id_M \times h)C(id_M \times h^{-1})$ to $M \times \{0, 1\} \times [0, 1]$ really is of the form $D \times id_{[0, 1]} | M \times \{0, 1\} \times [0, 1]$, where D is defined, as illustrated above, by $(id_M \times h)C(id_M \times h^{-1}) | M \times [0, 1] = M \times [0, 1] \times 0$.

When $\mathcal{A} = \mathcal{D}iff$, the proof is similar, only now we must take precautions to insure that the result is C^∞ where it is supposed to be. Notice that we need not worry about the behavior of h inside the shaded areas, because the resulting map $(id_M \times h^{-1})(D \times id_{[0, 1]})(id_M \times h)$ is independent of the parameters (r, s) in each of these areas. This follows from the assumptions of stationariness. We leave the construction of an appropriately smooth h, at this point, to the reader.

Actually, when $\mathcal{B} = \mathcal{H}$, it remains to prove that the map we construct belongs to

$$(\mathcal{H}, \mathcal{A})(M \times S,\ N \times S \text{ rel } X \times S).$$

Clearly, it does belong to the larger set

$$(\mathcal{H}', \mathcal{A})(M \times S,\ N \times S, \text{ rel } X \times S).$$

That is, the map is an $(\mathcal{H}', \mathcal{A})$-concordance rel $X \times [0, 1]$ between maps in

$$(\mathcal{H}, \mathcal{A})(M \times [0, 1],\ N \times [0, 1] \text{rel } X \times [0, 1]).$$

The desired result now follows from 3.1.8. (Note that this argument is not circular. Lemma 3.1.8 requires 3.1.9 in the special case $\mathcal{B} = \mathcal{H}'$, which has been proved above independently of 3.1.8.) Q.E.D.

<u>Proof of</u> 3.1.8:

Every $(\mathcal{H}, \mathcal{A})$-concordance is, by definition, an $(\mathcal{H}', \mathcal{A})$-concordance. We prove the converse. Suppose that H is an $(\mathcal{H}', \mathcal{A})$-concordance rel X between maps h_0 and $h_1 \in (\mathcal{H}, \mathcal{A})(M, N \text{ rel } X)$. Clearly, we may suppose that H is stationary within $\frac{1}{3}$ of $\{0, 1\}$. We apply Sublemma 3.1.9 to conclude that there exists an $(\mathcal{H}', \mathcal{A})$-concordance rel $X \times [0, 1]$ between H and $h_1 \times id_{[0, 1]}$. Since this latter map belongs to

$$(\mathcal{H}, \mathcal{A})(M \times [0, 1], N \times [0, 1] \text{ rel } X \times [0, 1]),$$

and this semigroup is, by definition, closed under $(\mathcal{H}', \mathcal{A})$-concordance rel $X \times [0, 1]$, the concordance H belongs to it. That is, H is an $(\mathcal{H}, \mathcal{A})$-concordance rel X. Q. E. D.

§3.2 THE GROUPS $\pi_i(\mathcal{A}; M \text{ rel } X)$ AND $\pi_{i+1}(\mathcal{B}, \mathcal{A}; M \text{ rel } X)$.

Throughout this chapter, $(\mathcal{B}, \mathcal{A})$ is a pair of geometric categories as in 3.1.1, M is a manifold in $\mathcal{A}$ with empty boundary, and X is a closed subset of M.

3.2.1 Some standard objects: Euclidean space of dimension i will be denoted by $\mathbb{R}^i$ and its standard coordinates by $x_1, x_2, \ldots, x_i$. We denote by $\mathbb{R}^i_+ \subseteq \mathbb{R}^i$ the half-space defined by $x_i \geq 0$ and by $\mathbb{R}^i_-$ the half-space defined by $x_i \leq 0$. The closed unit interval is, of course, $[0, 1]$. These are all endowed with their usual C^∞, PL, and topological structures, so that they are all objects in each geometric category.

3.2.2 Some notation:

For all $i \geq 0$, we introduce the following abbreviations:

$$\mathcal{A}_i(M \text{ rel } X) = \mathcal{A}(M \times \mathbb{R}^i \text{ rel } X \times \mathbb{R}^i)$$

$$(\mathcal{B}, \mathcal{A})_{i+1}(M \text{ rel } X) = (\mathcal{B}, \mathcal{A})(M \times \mathbb{R}^{i+1}_+, M \times \mathbb{R}^i \text{ rel } X \times \mathbb{R}^{i+1}_+)$$

and

$$\pi_i(\mathcal{A}; M \text{ rel } X) = \pi_0(\mathcal{A}_i(M \text{ rel } X))$$

$$\pi_{i+1}(\mathcal{B}, \mathcal{A}; M \text{ rel } X) = \pi_0((\mathcal{B}, \mathcal{A})_{i+1}(M \text{ rel } X)).$$

As shown in §3.1, these last two sets are groups with respect to map-composition.

Henceforth, to simplify notation and terminology still further, we shall suppress all references to X, except when essential. All our results still apply to arbitrary closed X, however.

3.2.3 Juxtaposition

It will be useful to single out a certain class of compositions of pairs of $\mathcal{A}$-maps or $(\mathcal{B}, \mathcal{A})$-maps, which we can roughly describe by saying that the maps in each pair have disjoint supports. We shall call such compositions juxtapositions. To obtain a precise uniform description, we introduce the following notions:

3.2.4 Definition: Let V be any of the manifolds $M \times \mathbb{R}^i$, $M \times \mathbb{R}^i \times [0, 1]$, $M \times \mathbb{R}_+^{i+1}$, $M \times \mathbb{R}_+^{i+1} \times [0, 1]$, $i \geq 1$, and let $y_j: V \longrightarrow \mathbb{R}$ or $\mathbb{R}_+$ be the standard projection onto the jth factor of $\mathbb{R}^i$ or $\mathbb{R}_+^{i+1}$. Let $f: V \longrightarrow V$ be any self-map with non-empty compact support, and define

$$\min_j f = \text{minimum } y_j(\text{supp } f)$$
$$\max_j f = \text{maximum } y_j(\text{supp } f)$$

Define $\min_j \text{id}_V = 1$, $\max_j \text{id}_V = -1$.

We call f j-positive (resp., j-negative) if $\min_j f > 0$ (resp., $\max_j f < 0$).

In the following lemma, we let $\mathcal{C}$ stand for $\mathcal{A}_i(M)$ or $(\mathcal{B}, \mathcal{A})_{i+1}(M)$, and let $\mathcal{C}$-concordance stand for $\mathcal{A}$-concordance or $(\mathcal{B}, \mathcal{A})$-concordance, respectively.

3.2.5 Lemma: Suppose that $1 \leq j \leq i$. Then:

a) Every map in $\mathcal{C}$ is $\mathcal{C}$-concordant to a j-positive (resp., j-negative) map in $\mathcal{C}$.

b) If two j-positive (resp., j-negative) maps in $\mathcal{C}$ are $\mathcal{C}$-concordant, then we may choose the $\mathcal{C}$-concordance to be j-positive (resp., j-negative).

Proof: Conjugate maps f in $\mathcal{C}$ by maps of the form $id_M \times T$, where T is a translation $\mathbb{R}^i \longrightarrow \mathbb{R}^i$ or $\mathbb{R}^{i+1}_+ \longrightarrow \mathbb{R}^{i+1}_+$ in the positive or negative direction of the jth coordinate. It is easy to choose T so that $(id_M \times T)f(id_M \times T^{-1}) = f^T$ is j-positive (resp., j-negative). Now, T is determined by a constant $a(T)$, the amount of translation in the jth coordinate. For $0 \leq t \leq 1$, let T_t denote the corresponding translation by $ta(T)$. Then, conjugation by $id_M \times T_t$, $0 \leq t \leq 1$, determines, after an easy adjustment to insure stationariness, a $\mathcal{C}$-concordance between f and f^T. This proves a). The proof of b) is similar. Q. E. D.

3.2.6 Definition: We still use $\mathcal{C}$ as in 3.2.5 above. If $f, g \in \mathcal{C}$, then a juxtaposition of f with g will be any composition f^+g^-, where f^+ is an i-positive map in $\mathcal{C}$ that is $\mathcal{C}$-concordant to f and g^- is an i-negative map in $\mathcal{C}$ that is $\mathcal{C}$-concordant to g.

3.2.7 Corollary: When juxtaposition is defined, that is when $i \geq 1$, the composition-induced group operations on $\pi_i(\mathcal{A}; M)$ and $\pi_{i+1}(\mathcal{B}, \mathcal{A}; M)$ are abelian.

Proof: Using the notation of 3.2.6, above, fg is $\mathcal{C}$-concordant to f^+g^- which equals g^-f^+ which is $\mathcal{C}$-concordant to gf. Q. E. D.

§3.3 EXACTNESS AND NATURALITY

Throughout this section, $(\mathcal{B}, \mathcal{A})$ and $(\bar{\mathcal{B}}, \bar{\mathcal{A}})$ will be pairs of geometric categories satisfying $(\mathcal{B}, \mathcal{A}) \leq (\bar{\mathcal{B}}, \bar{\mathcal{A}})$, by which we mean $\mathcal{B} < \bar{\mathcal{B}}$ or $\mathcal{B} = \bar{\mathcal{B}}$ and $\mathcal{A} < \bar{\mathcal{A}}$ or $\mathcal{A} = \bar{\mathcal{A}}$. M will be a manifold-without-boundary in $\mathcal{A}$. As before, we shall suppress mention of the closed set $X \subseteq M$, except when necessary, but we remind the reader that all our results and definitions hold for arbitrary closed X.

3.3.1 Definition: We define homomorphisms:

$$\pi_{i-1}(\mathcal{A};\, M) \xrightarrow{j} \pi_{i-1}(\mathcal{B};\, M)$$
$$\pi_i(\mathcal{B};\, M) \xrightarrow{k} \pi_i(\mathcal{B}, \mathcal{A}\,;\, M)$$
$$\pi_i(\mathcal{B}, \mathcal{A}\,;\, M) \xrightarrow{\partial} \pi_{i-1}(\mathcal{A};\, M)$$
$$\pi_i(\mathcal{B}, \mathcal{A}\,;\, M) \xrightarrow{\ell} \pi_i(\bar{\mathcal{B}}, \bar{\mathcal{A}}\,;\, M),$$

for all $i \geq 1$:

j is induced by the inclusion $\mathcal{A}_{i-1}(M) \subseteq \mathcal{B}_{i-1}(M)$.

k is induced by the restriction to $M \times \mathbb{R}^i_+$ of i-positive maps in $\mathcal{B}_i(M)$. By Lemma 3.2.5, k is well-defined.

∂ is induced by restriction to $M^n \times \mathbb{R}^{i-1} \subseteq M^n \times \mathbb{R}^i_+$.

ℓ is induced by the inclusion $(\mathcal{B}, \mathcal{A}\,)_i(M) \subseteq (\bar{\mathcal{B}}, \bar{\mathcal{A}}\,)_i(M)$.

3.3.2 Exactness Theorem: Let $i \geq 1$. The following sequence is exact:

$$\begin{array}{l} \ldots \to \pi_i(\mathcal{A};\, M) \xrightarrow{j} \pi_i(\mathcal{B};\, M) \\ \qquad\qquad\qquad\qquad\quad \big\downarrow k \\ \qquad\qquad\qquad\qquad\quad \pi_i(\mathcal{B}, \mathcal{A};\, M) \xrightarrow{\partial} \pi_{i-1}(\mathcal{A};\, M) \xrightarrow{j} \pi_{i-1}(\mathcal{B};\, M) \to \ldots \end{array}$$

We give a proof at the end of this section. We shall call this sequence the $(\mathcal{B}, \mathcal{A}\,)$-sequence for M (or for (M, X)).

We shall also use the letters j, k, ∂, ℓ somewhat ambiguously as generic notation for the kind of map described. Thus, because $\mathcal{A} \leq \bar{\mathcal{A}}$ and $\mathcal{B} \leq \bar{\mathcal{B}}$, we have

$$\pi_i(\mathcal{A};\, M) \xrightarrow{j} \pi_i(\bar{\mathcal{A}};\, M)$$

and

$$\pi_i(\mathcal{B};\, M) \xrightarrow{j} \pi_i(\bar{\mathcal{B}};\, M)\,.$$

3.3.3 First Naturality Theorem: The homomorphisms j above, together with the homomorphism ℓ of 3.3.1, form a map from the $(\mathcal{B}, \mathcal{A}\,)$ sequence for M to the $(\bar{\mathcal{B}}, \bar{\mathcal{A}}\,)$ sequence for M.

In other words, the homomorphisms j, j, and ℓ commute with the maps of the sequences. The proof is immediate.

3.3.4 Definition: Let M_α be a manifold in $\mathcal{Q} \neq \mathcal{H}$, $\partial M_\alpha = \emptyset$, and let X_α be a closed subset of M_α, $\alpha = 1, 2$.

A relative $\mathcal{Q}$-map $f: (M_1, X_1) \longrightarrow (M_2, X_2)$ is a continuous map of pairs for which

$$f^{-1} | M_2 - X_2$$

is a well-defined $\mathcal{Q}$-isomorphism from $M_2 - X_2$ onto an open subset $U \subseteq M_1 - X_1$.* If

$$U = M_1 - X_1,$$

then we call f a relative $\mathcal{Q}$-isomorphism.

Examples: a) If $X_1 \subseteq X_2 \subseteq M$, the inclusion

$$(M, X_1) \subseteq (M, X_2)$$

is a relative $\mathcal{Q}$-map.

b) Let $D^n_{\pm} = S^n \cap \mathbb{R}^{n+1}_{\pm}$, and let $\iota: D^n_- \longrightarrow M^n$ be an orientation-preserving, locally flat $\mathcal{Q}$-imbedding. Any map

$$(M^n, M^n - \text{int}\, \iota(D^n_-)) \longrightarrow (S^n, D^n_+)$$

which agrees with ι^{-1} on $\iota(D^n_-)$ is a relative $\mathcal{Q}$-isomorphism.

3.3.5 Definition: Let M_α, X_α, and $\mathcal{Q}$ be as in the above definition and let $f: (M_1, X_1) \longrightarrow (M_2, X_2)$ be a relative $\mathcal{Q}$-map. For every $h \in \mathcal{Q}_i(M_2 \text{ rel } X_2)$, $i \geq 0$, define

$$f^{\#}h \in \mathcal{Q}_i(M_1 \text{ rel } X_1)$$

by:

---

*Note that any map $f: (M_1, X_1) \longrightarrow (M_2, M_2)$ is a relative $\mathcal{Q}$-map for trivial reasons. Needless to say, this case is not of great interest.

$$f^{\#}h \mid U \times \mathbb{R}^i = (f^{-1} \times \mathrm{id}_{\mathbb{R}^i}) h (f \times \mathrm{id}_{\mathbb{R}^i}) \mid U \times \mathbb{R}^i$$

$$f^{\#}h \mid (M_1 - U) \times \mathbb{R}^i = \mathrm{id}_{M_1 \times \mathbb{R}^i} \mid (M_1 - U) \times \mathbb{R}^i$$

Here, U is the open set $f^{-1}(M_2 - X_2)$ described in Definition 3.3.4 above.*

A similar definition of $f^{\#}h$ applies when h is in $\mathcal{B}_i(M_2 \text{ rel } X_2)$ or in $(\mathcal{B}, \mathcal{A})_{i+1}(M_2 \text{ rel } X_2)$. Some verification is needed when $\mathcal{B} = \mathcal{H}$, however. We do this in the following proof.

3.3.6 Second Naturality Theorem: We use the above notation.

a) The association $h \longrightarrow f^{\#}h$ determines homomorphisms $f^{\#}$ on the concordance-homotopy-group-level which form a map from the $(\mathcal{B}, \mathcal{A})$ sequence for (M_2, X_2) to the $(\mathcal{B}, \mathcal{A})$-sequence for (M_1, X_1).

b) If f is a relative $\mathcal{A}$-isomorphism, then the homomorphisms $f^{\#}$ are isomorphisms.

Proof: a) We shall make use of the easily verified relation

$$(*) \qquad f^{\#}(gh) = (f^{\#}g)(f^{\#}h) .$$

That $f^{\#}h$ is a map of the required type is obvious except possibly when $h \in \mathcal{H}_i(M_2 \text{ rel } X_2)$ or $h \in (\mathcal{H}, \mathcal{A})_{i+1}(M_2 \text{ rel } X_2)$. We consider only the case $h \in \mathcal{H}_i(M_2 \text{ rel } X_2)$, the other being handled in exactly the same way. In this case, one shows first that $f^{\#}h$ belongs to the larger of the two sets

---

*In order to show that $f^{\#}h$ is well-defined, we use the relation $h(\mathrm{supp}\ h) \subseteq \mathrm{supp}\ h$ (cf. 3.1.5).

$$\mathcal{H}_i(M_1 \text{ rel } X_1) \subseteq \mathcal{H}'_i(M_1 \text{ rel } X_1). \; *$$

We shall show that it belongs to the smaller. Note that $\mathcal{H}_i(M_1 \text{ rel } X_1)$ is closed under $\mathcal{H}'$-concordance rel X_1 by definition.

Now choose $g \in \mathcal{H}_i(M_2 \text{ rel } X_2)$ and an $\mathcal{H}$-concordance rel X_2, $H: gh \sim id_{M_2 \times \mathbb{R}^i}$. Then, forming $f^{\#}H$ by analogy with the above, it is an $\mathcal{H}'$-concordance rel X_1 between $(f^{\#}g)(f^{\#}h) = f^{\#}(gh)$ and $f^{\#}(id_{M_2 \times \mathbb{R}^i}) = id_{M_1 \times \mathbb{R}^i}$. Similarly $(f^{\#}h)(f^{\#}g)$ is $\mathcal{H}'$-concordant rel X_1 to $id_{M_1 \times \mathbb{R}^i}$. Therefore,

$$f^{\#}h \in \mathcal{H}_i(M_1 \text{ rel } X_1),$$

as desired.

Similarly, by applying the $f^{\#}$-construction to concordances, one shows that $f^{\#}$ respects concordance classes. Here again special care must be taken when $\mathcal{H}$ is involved, for if H is an $\mathcal{H}$-concordance rel X_2, then $f^{\#}H$ is not obviously an $\mathcal{H}$-concordance rel X_1. However, it is an $\mathcal{H}'$-concordance between maps in $\mathcal{H}_i(M_1 \text{ rel } X_1)$, so that we may apply Lemma 3.1.9 to conclude that it is an $\mathcal{H}$-concordance. Similarly for $(\mathcal{H}, \mathcal{A})$-concordances.

Thus, $h \longrightarrow f^{\#}h$ determines <u>maps</u> on the concordance-homotopy-group level. Relation (*) above shows that these maps are homomorphisms,

---

* In other words, $f^{\#}h$ is: a) a homotopy equivalence $M_1 \times \mathbb{R}^i \to M_1 \times \mathbb{R}^i$, and b) it has compact support that avoids $X \times \mathbb{R}^i$. Assertion b) is clear. When $i \geq 1$, assertion a) follows, for example, from the fact that $f^{\#}h | M \times \{v\}$ is the identity for all $v \in \mathbb{R}^i$ of sufficiently large norm. We can prove assertion a) for all $i \geq 0$ in the same way that we show $f^{\#}h \in \mathcal{H}_i(M_1 \text{ rel } X_1)$, later in the above proof, or we can show that it induces isomorphisms of homology and homotopy and then apply Whitehead's theorem.

as required.

The verification that the homomorphisms commute with the maps of the $(\mathcal{B}, \mathcal{A})$-sequence is trivial.

b) If f is a relative $\mathcal{A}$-isomorphism, we can define, on the level of maps, an association inverse to $h \longrightarrow f^{\#}h$, which preserves compositions and concordance classes. Q. E. D.

3.3.6 *Corollary: Let* $\iota: D^n_- \longrightarrow M^n$ *be as in* Example b) *of* 3.3.4, *with* $X \cap \iota(\text{int } D^n_-) = \emptyset$. *Let* f *be the composite*

$$(M^n, X) \subseteq (M^n, M^n - \text{int } \iota(D^n_-)) \longrightarrow (S^n, D^n_+),$$

where the second map is as described in 3.3.4. *Then,*

a) $f^{\#}$ *takes the* $(\mathcal{B}, \mathcal{A})$ *sequence for* (S^n, D^n_+) *into the* $(\mathcal{B}, \mathcal{A})$ *sequence for* (M^n, X). *It is an isomorphism* if $X = M^n - \text{int } \iota(D^n_-)$

b) $f^{\#}$ *depends only on the path component of* $M^n - X$ *that contains* $\iota(D^n_-)$, *except possibly when* $\mathcal{A} = \mathcal{T}op$ *and* $n = 4$.

Proof: Statement a) is just a restatement of 3.3.5.

To prove statement b) when $n \neq 4$, or $n = 4$ and $\mathcal{A} \neq \mathcal{T}op$, observe that in these cases any two orientation-preserving, locally-flat $\mathcal{A}$-imbeddings $D^n \longrightarrow M^n - X$ that are homotopic (i.e., have images in the same path-component) are $\mathcal{A}$-ambient isotopic via a compactly supported isotopy. Thus, if we use these imbeddings to construct $f_1^{\#}, f_2^{\#}$, the ambient isotopy will provide us with concordances $f_1^{\#}h \sim f_2^{\#}h$, for all h. Q. E. D.

Remark: In §3.4, we show that

$$0 \approx \pi_i(\mathcal{T}op; S^n, D^n_+) \approx \pi_i(\mathcal{H}; S^n, D^n_+) \approx \pi_{i+1}(\mathcal{H}, \mathcal{T}op; S^n, D^n_+),$$

for all n and i. Therefore, the exception in statement b) may be deleted for trivial reasons.

In fact, the only non-trivial concordance-homotopy groups for

(S^n, D^n_+), in general, are

$$\pi_i(\mathcal{D}iff; S^n \operatorname{rel} D^n_+) \approx \pi_{i+1}(\mathcal{B}, \mathcal{D}iff; S^n \operatorname{rel} D^n_+)$$

(see 3.4.1).

Before we prove the Exactness Theorem, we state the following lemma. A proof will be given at the end of this section.

Let $c\colon M \times \mathbb{R}^{i-1} \times [0, 1] \longrightarrow M \times \mathbb{R}^i_+$ be the natural collar; that is, c is given by

$$c(p, (x_1, \ldots, x_{i-1}), t) = (p, (x_1, \ldots, x_{i-1}, t)).$$

3.3.7 <u>Lemma</u>: a) <u>Every</u> $f \in (\mathcal{B}, \mathcal{A})_i(M)$ <u>is</u> $(\mathcal{B}, \mathcal{A})$<u>-concordant to a map</u> g <u>satisfying</u>: (i) $\max_i g < \frac{1}{3}$; (ii) g <u>is stationary near</u> $M \times \mathbb{R}^{i-1}$ <u>with respect to</u> c.

b) <u>If</u> F <u>is a</u> $(\mathcal{B}, \mathcal{A})$<u>-concordance between maps</u> f_0 <u>and</u> f_1 <u>in</u> $(\mathcal{B}, \mathcal{A})_i(M)$ <u>that satisfy</u> (i) <u>and</u> (ii) <u>above, then there exists a</u> $(\mathcal{B}, \mathcal{A})$<u>-concordance</u> G <u>between</u> f_0 <u>and</u> f_1 <u>such that</u>: (i) $\max_i G < \frac{1}{3}$; (ii) G <u>is stationary near</u> $M \times \mathbb{R}^{i-1} \times [0, 1]$ <u>with respect to the natural collar</u>

$$d\colon (M \times \mathbb{R}^{i-1} \times [0, 1]) \times [0, 1] \longrightarrow M \times \mathbb{R}^i_+ \times [0, 1].$$

(iii) $G|M \times \mathbb{R}^{i-1} \times [0, 1] = F|M \times \mathbb{R}^{i-1} \times [0, 1]$.

Here, d is given by

$$d((p, x, s), t) = (c(p, x, t), s).$$

<u>Proof of the Exactness Theorem:</u>

We shall prove that

$$\text{kernel } k \subseteq \text{image } j, \ \text{kernel } \partial \subseteq \text{image } k, \ kj = 0.$$

The remaining three relations have similar but easier proofs. Each of the above relations will be verified by, <u>first</u>, reducing the verification to the problem of extending a certain map H on

$$W = (M \times \mathbb{R}^i_+ \times 0) \cup (M \times \mathbb{R}^{i-1} \times [0, 1]) \cup (M \times \mathbb{R}^i_+ \times 1)$$

to a $(\mathcal{B}, \mathcal{A})$-concordance on $M \times \mathbb{R}^i_+$, and, secondly, using Lemma 3.1.9 (or a similar fact) to conclude that such an extension exists. Reference to X is suppressed, as usual, for expository convenience.

a) kernel $k \subseteq$ image j.

Let f be an i-positive map (see 3.2.4) in $\mathcal{B}_i(M)$ such that there exists a $(\mathcal{B}, \mathcal{A})$-concordance F between $F_0 = \mathrm{id}_{M \times \mathbb{R}^i_+}$ and $F_1 = f | M \times \mathbb{R}^i_+$. We must show that f is $\mathcal{B}$-concordant to a map in $\mathcal{A}_i(M)$.

We define H on W by the following scheme:

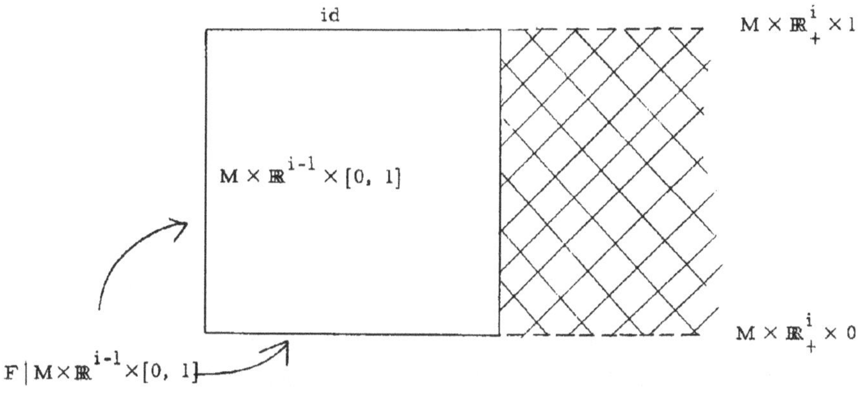

Here, for the lower arrow, we are identifying

$$\{(p, x_1, \ldots, x_i) \mid 0 \le x_i \le 1\}$$

with the cylinder $M \times \mathbb{R}^{i-1} \times [0, 1]$ in the obvious way. Now, extend H to the cross-hatched area via the identity map. We are, then, exactly in the situation of Lemma 3.1.9,* and so, we may fill-in the remaining region with a $\mathcal{B}$-map that is stationary near W with respect to the given product structure and the map d (of Lemma 3.3.7). Call the result K, and let $-K$ be obtained from K by conjugating it with the isomorphism

$$M \times \mathbb{R}^i_- \times [0, 1] \longrightarrow M \times \mathbb{R}^i_+ \times [0, 1]$$

that simply reverses the sign of x_i. Define a map G by the following scheme:

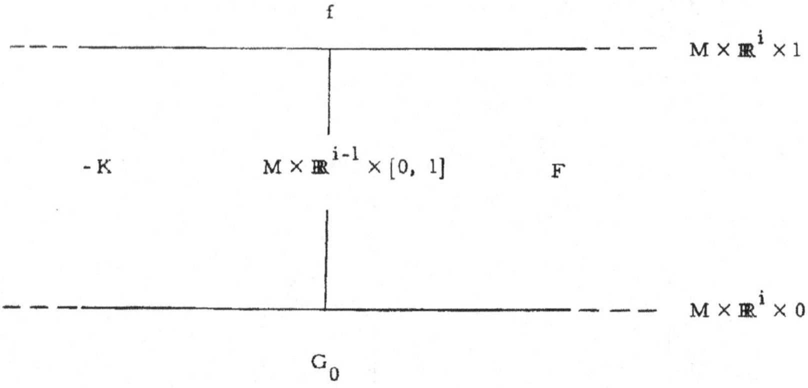

We may suppose, without loss of generality, that F is stationary near $M \times \mathbb{R}^{i-1} \times [0, 1]$ (Lemma 3.3.7), so that G is an $\mathcal{A}$-map in a neighborhood of $M \times \mathbb{R}^{i-1} \times [0, 1]$. In particular, $G_0 \in \mathcal{A}_i(M)$.

When $\mathcal{B} \neq \mathcal{H}$, G is clearly a $\mathcal{B}$-concordance between f and G_0, as desired. When $\mathcal{B} = \mathcal{H}$, we see immediately that G is an $\mathcal{H}'$-concordance between f and G_0 (see 3.1.7). Since both f and G_0 belong to $\mathcal{H}(M)$, Lemma 3.1.8 implies that they are

---

* See footnote for part b).

$\mathcal{H}$-concordant. This completes the proof of a).

b) $kj = 0$

Let f be an i-positive map in $\mathcal{A}_i(M)$. We must show that $f_+ = f|M \times \mathbb{R}^i_+$ is $(\mathcal{B}, \mathcal{A})$-concordant to $id_{M\times\mathbb{R}^i_+}$. By Lemma 3.3.7, we may assume that $\max_i f_+ < 1$. We define H on W by the following scheme

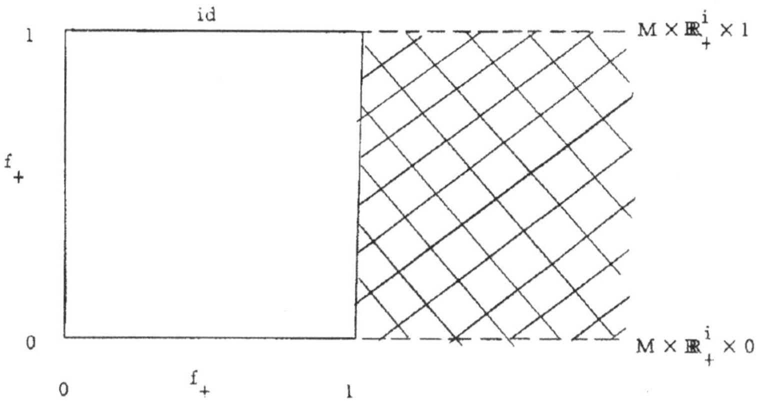

and fill in the cross-hatched area by the identity map. Again we are in the situation of Lemma 3.1.9* so that an extension F exists, stationary near W.

When $\mathcal{B} \neq \mathcal{H}$, F is obviously a $(\mathcal{B}, \mathcal{A})$-concordance. When $\mathcal{B} = \mathcal{H}$, we use the argument of part a).

c) kernel $\partial \subseteq$ image k

Choose $f \in (\mathcal{B}, \mathcal{A})_i(M)$ such that there is an $\mathcal{A}$-concordance

---

*Actually, Lemma 3.1.9 requires that f_+ be stationary within $\frac{1}{3}$ of $\{0, 1\}$. The number $\frac{1}{3}$ is clearly inessential here, however, and was chosen only for expository convenience.

F between $F_0 = \mathrm{id}_{M \times \mathbb{R}^{i-1}}$ and $F_1 = f | M \times \mathbb{R}^{i-1}$. We must show that f is $(\mathcal{B}, \mathcal{A})$-concordant to the restriction of an i-positive map in $\mathcal{B}_i(M)$.

By 3.3.7, we may suppose that $\max_i f < \frac{1}{3}$ and that f is stationary near $M \times \mathbb{R}^{i-1}$. It is easy to see that we may also assume F to be stationary within $\frac{1}{3}$ of $\{0, 1\}$. We define H on W by the following scheme

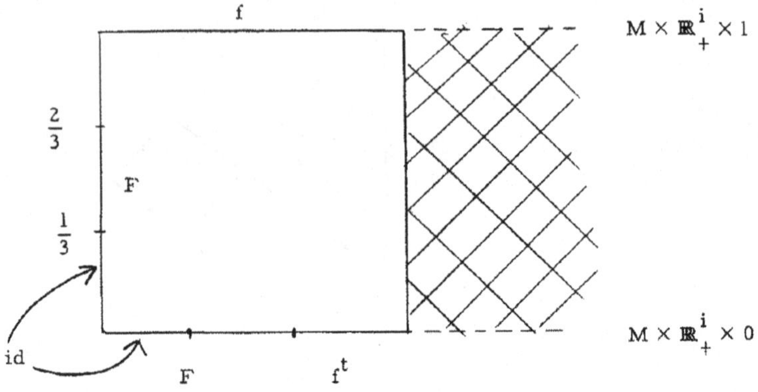

Here, f^t is the map obtained by translating f 2/3 of a unit in the positive x_i direction (i.e., conjugating with $\mathrm{id}_M \times \lambda$, where λ is this translation). Fill in the cross-hatched area with the identity. We are now in a situation analogous to that of Lemma 3.1.9, and a similar proof shows that there is an extension G of H, stationary near W. When $\mathcal{B} \neq \mathcal{H}$, G is obviously a $(\mathcal{B}, \mathcal{A})$-concordance between f and the restriction of an i-positive map in $\mathcal{B}_i(M)$. When $\mathcal{B} = \mathcal{H}$, this requires a few more words.

First, we conclude that G is an $(\mathcal{H}', \mathcal{A})$-concordance between f and $G_0 \in (\mathcal{H}', \mathcal{A})_i(M)$. Since $(\mathcal{H}, \mathcal{A})_i(M)$ is closed under $(\mathcal{H}', \mathcal{A})$-concordance, it follows that $G_0 \in (\mathcal{H}, \mathcal{A})_i(M)$, and Lemma 3.1.8 implies that G is an $(\mathcal{H}, \mathcal{A})$-concordance. We have shown, therefore, that f is $(\mathcal{H}, \mathcal{A})$-concordant to the restriction of an i-positive map $\bar{f} \in \mathcal{H}_i^!(M)$ --i.e., $\bar{f}|M \times \mathbb{R}^i_+ \equiv \bar{f}_+ = G_0$.

Next, we may apply the same result to any $(\mathcal{H}, \mathcal{A})$-inverse (up to $(\mathcal{H}, \mathcal{A})$-concordance) g of f, obtaining $\bar{g} \in \mathcal{H}_i^!(M)$. It follows that $(\bar{f}\bar{g})_+$ is $(\mathcal{H}, \mathcal{A})$-concordant to $id_{M \times \mathbb{R}^i_+}$. By the argument of a), above, it follows that $\bar{f}\bar{g}$ is $\mathcal{H}'$-concordant to a map $h \in \mathcal{A}_i(M) \subseteq \mathcal{H}_i(M)$. Therefore, $\bar{f}(\bar{g}h^{-1})$ is $\mathcal{H}'$-concordant to $id_{M \times \mathbb{R}^i}$. Similarly, $(k^{-1}\bar{g})\bar{f}$ is $\mathcal{H}'$-concordant to $id_{M \times \mathbb{R}^i}$, for some $k \in \mathcal{A}_i(M)$. It follows that $\bar{f}$ represents a unit in $\mathcal{H}_i^!(M)$. That is, $\bar{f} \in \mathcal{H}_i(M)$, as desired. Q. E. D.

<u>Proof of Lemma</u> 3.3.7

a) Suppose that $f \in (\mathcal{B}, \mathcal{A})_i(M)$ is stationary near $M \times \mathbb{R}^{i-1}$ with respect to c, and let $\alpha: \mathbb{R}_+ \longrightarrow \mathbb{R}_+$ be given by $\alpha(x) = x/3a$, where a is an arbitrary but fixed number strictly greater than $\max_i f$. Let $T = id_{M \times \mathbb{R}^{i-1}} \times \alpha$, and define $g = T^{-1}fT : M \times \mathbb{R}^i_+ \longrightarrow M \times \mathbb{R}^i_+$. Then, g satisfies (i) and (ii). To show that g is $(\mathcal{B}, \mathcal{A})$-concordant to f, construct an $\mathcal{A}$-isotopy T_t between T and $id_{M \times \mathbb{R}^i}$. A "straight-line isotopy" is a good first approximation for T_t, but it must be modified in two ways. First, it must be adjusted to be stationary near $\{0, 1\}$; this is elementary. Secondly, when $\mathcal{A} = \mathcal{PL}$, we either C^1-approximate the (non-PL) straight-line isotopy by one that is PL, using PD triangulation theory, or we proceed directly by constructing a PL isotopy between α and $id_{\mathbb{R}_+}$. This last step is left as an exercise for the reader. Then, $T_t^{-1}fT_t$ gives the desired concordance.

To complete the proof of a), then, it suffices to find a $g \in (\mathcal{B}, \mathcal{A})_i(M)$, stationary near $M \times \mathbb{R}^{i-1}$ with respect to c, that is $(\mathcal{B}, \mathcal{A})$-concordant to some given $f \in (\mathcal{B}, \mathcal{A})_i(M)$.

We base the proof on the uniqueness theorem for collar neighborhoods (see [16], [5] for the $\mathcal{PL}$ and $\mathcal{Top}$ cases). Let $f_0 = f|M \times \mathbb{R}^{i-1}$. Recall that f is an $\mathcal{A}$-isomorphism on a neighborhood of $M \times \mathbb{R}^{i-1} \subset M \times \mathbb{R}^i_+$, so that, since supp f is compact, we may find an $\varepsilon > 0$ for which

$$f|\{(p, x) \in M \times \mathbb{R}^i_+ | x_i < 3\varepsilon\}$$

is an $\mathcal{A}$-isomorphism. Then, both $c|M \times \mathbb{R}^{i-1} \times [0, 2\varepsilon]$ and

$$c' = f \circ c \circ (f_0 \times id_{[0, 2\varepsilon]})^{-1}$$

are $\mathcal{A}$-collars for $M \times \mathbb{R}^{i-1} \subseteq M \times \mathbb{R}^i_+$ which coincide outside some compact set K containing supp $f_0 \times [0, 2\varepsilon]$. (When the closed set $X \neq \emptyset$, K may be chosen to avoid $X \times \mathbb{R}^{i-1} \times [0, 2\varepsilon]$.) It follows from the collar uniqueness theorems that there is an ambient $\mathcal{A}$-isotopy G such that:

(1) supp G is compact (and avoids $X \times \mathbb{R}^i_+ \times [0, 1]$).

(2) $G(p, x, t) = (p, x, t)$, for $(p, x, t) \in M \times \mathbb{R}^{i-1} \times [0, 1]$.

(3) $G_1 \circ c'|M \times \mathbb{R}^{i-1} \times [0, \varepsilon] = c|M \times \mathbb{R}^{i-1} \times [0, \varepsilon]$.

The desired $(\mathcal{B}, \mathcal{A})$-concordance, then, is

$$F = G \circ (f \times id_{[0, 1]}).$$

b) If we find a G that satisfies b) (ii) and (iii), then we obtain b) (i) just as in a) above. We omit this step, therefore, and proceed to find a G satisfying (ii) and (iii). Suppose that F is stationary within $2\varepsilon > 0$ of $\{0, 1\}$. Let $F_0 = F|M \times \mathbb{R}^{i-1} \times (\varepsilon, 1-\varepsilon)$ and choose $\delta > 0$ so that

$$F|\{(p, x, t) \in M \times \mathbb{R}^i_+ \times (\varepsilon, 1-\varepsilon) | x_i < 3\delta\}$$

is an $\mathcal{A}$-isomorphism. Then, $d|(M \times \mathbb{R}^{i-1} \times (\varepsilon, 1-\varepsilon)) \times [0, 2\delta]$ and

$$d' = F \circ d \circ (F_0 \times id_{[0, 2\delta]})^{-1}$$

are $\mathcal{A}$-collars for $M \times \mathbb{R}^{i-1} \times (\varepsilon, 1-\varepsilon) \subseteq M \times \mathbb{R}^i_+ \times [0, 1]$ which coincide outside $M \times \mathbb{R}^{i-1} \times [2\varepsilon, 1-2\varepsilon] \times [0, 2\delta]$. Thus, there exists an ambient $\mathcal{A}$-isotopy G_t of $M \times \mathbb{R}^i_+ \times [0, 1]$, with compact support that avoids $M \times \mathbb{R}^i_+ \times [0, \varepsilon]$ and $M \times \mathbb{R}^i_+ \times [1-\varepsilon, 1]$, such that $G_1 \circ d' | M \times \mathbb{R}^{i-1} \times (\varepsilon, 1-\varepsilon) \times [0, \delta] = d | M \times \mathbb{R}^{i-1} \times (\varepsilon, 1-\varepsilon) \times [0, \delta]$. Let $G = G_1 \circ F$. Q.E.D.

The argument in b) is easily adapted to the case $X \neq \emptyset$.

§3.4 SPECIAL COMPUTATIONS OF $\pi_i(\mathcal{A};M \text{ rel } X)$

3.4.1 Proposition:

$$\pi_i(\mathcal{A};S^n \text{ rel } D^n_+) = \begin{cases} 0 & , \mathcal{A} \neq \mathcal{D}iff \\ \Gamma^{n+i+1} & , \mathcal{A} = \mathcal{D}iff \end{cases}$$

where Γ^{n+i+1} is the Kervaire-Milnor-Smale group of exotic $(n+i+1)$-spheres.

Proof: When $\mathcal{A} = \mathcal{D}iff$ this result was proved in Chapter 2, [1]. When $\mathcal{A} \neq \mathcal{D}iff$, we use the Alexander "trick" or Alexander isotopy (e.g., see [22]). Note that for the trick to work, we must have, for $h \in \mathcal{A}_i(S^n \text{ rel } D^n_+)$,

$$h(\text{supp}h) \subseteq \text{supp}h.$$

When $\mathcal{A} \neq \mathcal{H}$, this is immediate, because h is a homeomorphism. When $\mathcal{A} = \mathcal{H}$, it is also true but requires a check (cf. 3.1.5). Q.E.D.

In §3.5, we state our Classification Theorem. This reduces the problem of calculating $\pi_{i+1}(\mathcal{B},\mathcal{A};M \text{ rel } X)$ to problems in homotopy theory and surgery theory (see Theorem 3.5.5). In this section, we describe certain circumstances under which $\pi_i(\mathcal{H};M \text{ rel } X)$ can be computed by standard algebraic-topological techniques. Combining these two facts with the exactness of the $(\mathcal{H},\mathcal{A})$-sequence for (M, X), we see that, under certain circumstances, we can obtain substantial information about $\pi_i(\mathcal{A};M \text{ rel } X)$ via algebraic topology and surgery theory. We present some applications of this in 3.5.7, 3.5.9.

3.4.2 Proposition: Suppose that M is a closed manifold and that the closed set X is a subcomplex of some triangulation of an open subset of M. Suppose also that M-X admits the structure of a homotopy-associative H-space with homotopy-identity and homotopy-inverses. Then:

$$\pi_i(\mathcal{H}; M \text{ rel } X) \approx [\Sigma^i(M/X), M\text{-}X], \quad i \geq 1.^*$$

Setting $M = S^n$ and $X = D^n_+$, we recapture part of 3.4.1. We give a proof of 3.4.2 in Appendix D, together with a number of corollaries. We shall not make essential use of 3.4.2 in our main development, however. Instead, we use the following easy result.

3.4.3 Proposition: If the closed, triangulable manifold M is a $K(\pi, 1)$, then

$$\pi_i(\mathcal{H}; M) = 0, \quad i > 1.$$

Proof: It is clear that every class in $\pi_i(\mathcal{H}; M)$ has a representative $f \in \mathcal{H}_i(M)$ with

$$\text{supp } f \subsetneq \text{int } (M \times D^i),$$

where $D^i \subsetneq \mathbb{R}^i$ is the unit ball. Let

$$g = f | M \times D^i : M \times D^i \to M \times D^i.$$

---

*If A and B are spaces with basepoint, then [A, B] denotes the set of based homotopy classes of based maps $A \to B$. Choose arbitrary basepoints for $\Sigma^i(M/X)$ and M-X.

We shall show that g is homotopic rel $(M \times \partial D^i)$ to $id_{M \times D^i}$. Extend this homotopy by the identity to all of $M \times \mathbb{R}^i$. The extension yields an $\mathcal{H}$-concordance between f and $id_{M \times \mathbb{R}^i}$, as required.

First, let us triangulate M and give $M \times D^i$ a triangulation compatible with the PL product structure. Suppose then that we have constructed a homotopy rel $(M \times \partial D^i)$ over the 1-skeleton of $M \times D^i$, between g and id. The obstructions to further extension lie in cohomology groups

$$H^j(M \times D^i, M \times \partial D^i; \{\pi_j(M \times D^i)\}), \quad j \geq 2,$$

where $\{\pi_j(M \times D^i)\}$ is the natural local coefficient system of homotopy groups. Since $M \times D^i$ is a $K(\pi, 1)$, this system is the trivial one so that obstructions vanish. Therefore, the homotopy extends over all of $M \times D^i$.

It remains to obtain the desired homotopy over the 1-skeleton. Note that the pointed set

$$\pi_1(M \times D^i, M \times \partial D^i)$$

is trivial when $i > 1$. (It consists of two elements when $i = 1$.) It follows that the inclusion of 1-skeletons

$$e_0 : ((M \times D^i)_1, (M \times \partial D^i)_1) \to (M \times D^i, M \times \partial D^i)$$

is homotopic rel $(M \times \partial D^i)_1$ to a map e_1 satisfying

$$e_1((M \times D^i)_1) \subseteq M \times \partial D^i.$$

Recall that $g/M \times \partial D^i$ is the inclusion $M \times \partial D^i \subseteq M \times D^i$. Therefore, we have the following homotopies rel $(M \times \partial D^i)_1$:

$$g \circ e_0 \sim g \circ e_1 = e_1 \sim e_0 = \mathrm{id} \circ e_0.$$

That is, $\quad g/(M \times D^i)_1 \sim \mathrm{id}/(M \times D^i)_1 \quad \mathrm{rel}\ (M \times \partial D^i)_1,$

as desired. Q. E. D.

Remarks: (1) When π is abelian, M has the structure of an H-space (homotopy associative, etc.), so that in this case, we can derive 3.4.3 from 3.4.2. Note that the only possible abelian π, here, are the free-abelian ones, and M is just a homotopy torus.

(2) The hypothesis of triangulability in 3.4.3 can be removed in the following way. First, Kirby and Siebenmann [18] have shown that there exist homotopy equivalences of pairs

$$(M \times D^i, M \times \partial D^i) \underset{k}{\overset{h}{\rightleftarrows}} (K, L), \ k \circ h \sim \mathrm{id},$$

where K and L are finite simplicial complexes. Let

$$W = M \times D^i \times [0,1]$$

$$V = \partial W$$

$$L_0 = (K \times 0) \cup (L \times [0,1]) \cup (K \times 1).$$

Then,

$$(W, V) \underset{k \times 1}{\overset{h \times 1}{\rightleftarrows}} (K \times [0,1], L_0)$$

are homotopy equivalences of pairs.

We now prove, exactly as in 3.4.3 above, that $g \circ k \sim k$ rel L. Define G on V to be g on $M \times D^i \times 0$, and to send (p, x, t) to (p, x) elsewhere. We must show that G extends over W. But, we have just seen that

$$G \circ (k \times 1)/L_0$$

extends over $K \times [0,1]$. Therefore, $G \circ (k \times 1) \circ (h \times 1)/V$ extends over W. Since G is homotopic to $G \circ (kh \times 1)$, and since W is an ANR, the homotopy extension theorem now yields the desired result.

§3.5. A CLASSIFICATION THEOREM FOR $\pi_i(\mathcal{B}, \mathcal{A}; M \text{ rel } X)$

Throughout this section, $(\mathcal{B}, \mathcal{A})$ will be a pair of geometric categories, M will be a closed manifold in $\mathcal{A}$, and X will be a closed subset of M. The $(\mathcal{PL}, \mathcal{D}iff)$ case, which is essentially the same, will be treated in Appendix C.

For our classification theorem, we shall impose a regularity condition on X.

3.5.1 Definition: The closed subset $X \subseteq M$ is called regular if M-X is connected and if there exists a (possibly empty) compact, codimension-0, locally-flat $\mathcal{A}$-submanifold V of M with the following properties:

a) $X \subseteq$ interior V.

b) There exists an $\mathcal{A}$-isomorphism $h : \partial V \times [0, \infty) \to V - X$ with $h(x, 0) = x$, for all $x \in \partial V$.

Such a V will be called a regularizing manifold for X.

3.5.2 Examples: For each of the categories $\mathcal{A} = \mathcal{D}iff$, $\mathcal{PL}$, $\mathcal{T}op$ and for each manifold N in $\mathcal{A}$, we can define the notion of an $\mathcal{A}$-triangulation of N.* It is a homeomorphism $K \to N$, where K is a combinatorial manifold, whose restriction to each closed simplex in K is a C^∞, PL or topological imbedding, as the case may be.

Let U be an open subset of M admitting $\mathcal{A}$-triangulations, and let X be a compact polyhedron relative to one such triangulation, such that M-X is connected. Then X is regular in M. A regularizing manifold in the $\mathcal{PL}$ and $\mathcal{T}op$ cases is provided by a regular neighborhood of X in U. In the $\mathcal{D}iff$ case, one obtains V by smoothing this neighborhood via Hirsch [13].

In the $\mathcal{PL}$ and $\mathcal{D}iff$ cases, every U admits $\mathcal{A}$-triangulations (cf. Appendix C). In the $\mathcal{T}op$ case, Kirby and Siebenmann have given counterexamples [20]. Let X be such a (connected) counterexample. Then, $X = X \times * \subseteq X \times S^1$ is regular but it is not contained in the examples of the preceding paragraph.

---

*When $\mathcal{A} = \mathcal{D}iff$, this is sometimes called a PD triangulation. See Appendix C.

Note that every compact, codimension-0, locally-flat $\mathcal{A}$-submanifold $X \subseteq M$ with $M-X$ connected is regular. In particular, regularizing manifolds for regular X are themselves regular.

The following result shows that we may replace X by any of its regularizing manifolds.

3.5.3 Proposition: *Let V be a regularizing manifold for X. Then, the relative $\mathcal{A}$-map*

$$(M, X) \subseteq (M, V)$$

determines an isomorphism of exact $(\mathcal{B},\mathcal{A})$-sequences. In particular

$$\pi_i(\mathcal{B},\mathcal{A}; M \text{ rel } V) \approx \pi_i(\mathcal{B},\mathcal{A}; M \text{ rel } X).$$

Proof: We prove only the last statement; the rest follows in exactly the same way. Note that ∂V can be empty only when $X = V$, in which case the result is trivial. Therefore, suppose that $\partial V \neq \phi$, and let $h : \partial V \times [0,\infty) \to V-X$ be an $\mathcal{A}$-isomorphism with $h(x, 0) = x$, for all $x \in \partial V$, as provided by 3.5.1. For all $t \in [0,\infty)$, define

$$V_t = V - h(\partial V \times [0, t)).$$

Note that if $0 < s < t$, there exists an ambient $\mathcal{A}$-isotopy H_r of M such that it is stationary near $\{0,1\}$ and

$$H_r | V_t \quad \text{is the identity, } 0 \leq r \leq 1,$$

$$H_1(V_s) = V.$$

We use this to prove surjectivity.

Thus, let $f \in (\mathcal{B},\mathcal{Q})_i(M \text{ rel } X)$, and let S be the projection into M of $\text{supp } f \subseteq M \times \mathbb{R}^i_+$. Then, $S \cap (V-X)$ is a compact subset of $V-X$, so that, for sufficiently large s,

$$S \cap V_s = \phi.$$

It follows that, for such s,

$$f \in (\mathcal{B},\mathcal{Q})_i(M \text{ rel } V_s).$$

Let $t = s+1$ and construct the above-described isotopy H. Define $K : M \times \mathbb{R}^i_+ \times [0,1] \to M \times \mathbb{R}^i_+ \times [0,1]$ by $K(x,y,r) = (H_r(x), y, r)$. Then, $K \circ (f \times id_{[0,1]}) \circ K^{-1}$ is a $(\mathcal{B},\mathcal{Q})$-concordance rel X between f and $K_1 \circ f \circ K_1^{-1} \in (\mathcal{B},\mathcal{Q})_i(M \text{ rel } V)$.

To prove injectivity, we use the following easily checked fact: for every $t > 0$ and every open set $U \supset V$, there exists an $\mathcal{Q}$-isomorphism $g : M \to M$ such that

(1) $g(x) = x$, if $x \notin U$

(2) $g(V) = V_t$.

Now, let f_0 and f_1 be maps in $(\mathcal{B},\mathcal{Q})_i(M \text{ rel } V)$, and let H be a $(\mathcal{B},\mathcal{Q})$-concordance rel X between them. Then, just as before, for sufficiently large t, H is a $(\mathcal{B},\mathcal{Q})$-concordance rel V_t. Let $S_i \subseteq M$ be the projection of supp f_i, $i = 0,1$, and choose the open set $U \supset V$ such that $(S_0 \cup S_1) \cap \text{closure } U = \phi$. Construct g as above, and let

$G = g \times id_{\mathbb{R}^i_+ \times [0,1]}$. Then, $G^{-1}HG$ is a $(\mathcal{B},\mathcal{A})$-concordance rel V between f_0 and f_1. Q. E. D.

We now introduce some terminology.

3.5.4. Terminology and Remarks.

Let BSO, BSPL, BSTOP and BSG denote the classifying spaces for stable oriented vector bundles, stable oriented PL microbundles, stable oriented topological microbundles, and stable oriented spherical fibre spaces, respectively. Let SO, SPL, STOP and SG be the corresponding loop spaces. For any geometric category $\mathcal{A}$, $\mathcal{B}$, etc., we shall denote by the associated Roman letter A, B, etc., the loop space corresponding to it under the obvious rule:

$$\mathcal{D}iff \longleftrightarrow SO$$
$$\mathcal{PL} \longleftrightarrow SPL$$
$$\mathcal{T}op \longleftrightarrow STOP$$
$$\mathcal{H} \longleftrightarrow SG.$$

If $\mathcal{A} < \mathcal{B}$, there is a corresponding natural map of classifying spaces whose fibre we denote by B/A. If $(\mathcal{B},\mathcal{A}) \leq (\overline{\mathcal{B}},\overline{\mathcal{A}})$, then there is a corresponding based map $B/A \to \overline{B}/\overline{A}$, basepoints being fixed once and for all. For every Y with basepoint, this induces a map

$$[Y, B/A] \to [Y, \overline{B}/\overline{A}],$$

which we call a forgetful map. We also apply the term "forgetful" to the change-of-categories homomorphism

$$\ell : \pi_i(\mathcal{B},\mathcal{A};M \text{ rel } X) \to \pi_i(\overline{\mathcal{B}},\overline{\mathcal{A}};M \text{ rel } X)$$

defined in 3.3.1.

Recall that a relative $\mathcal{A}$-map (see 3.3.4, 3.3.5)

$$f : (M_1, X_1) \to (M_2, X_2)$$

determines a homomorphism

$$f^{\#} : \pi_i(\mathcal{B},\mathcal{A};M_2 \text{ rel } X_2) \to \pi_i(\mathcal{B},\mathcal{A};M_1 \text{ rel } X_1).$$

It also determines a map

$$M_1/X_1 \to M_2/X_2$$

which, for every based Z induces

$$f^{*} : [M_2/X_2, Z] \to [M_1/X_1, Z].$$

When $X = \phi$, M/X is understood to equal $M \cup$ basepoint.

3.5.5. <u>Classification Theorem</u>:* <u>Suppose that</u> X <u>is regular in</u> M. <u>Then, for each</u> $i \geq 1$, <u>there is a homomorphism</u>

---

*We repeat our remark of 3.14. All of the results in this section that we state for pairs $(\mathcal{B},\mathcal{A})$ will hold for the pair $(\mathcal{PL}, \mathcal{D}iff)$ unless we explicitly exclude $\mathcal{B} = \mathcal{PL}$ or $\mathcal{A} = \mathcal{D}iff$. Thus, for example, 3.5.5 a), b) applies to $(\mathcal{PL}, \mathcal{D}iff)$. Proofs for this case are given in Appendix C, however.

$$\mathcal{K}_i = \mathcal{K}_i(\mathcal{B},\mathcal{A};M \text{ rel } X) : \pi_i(\mathcal{B},\mathcal{A};M \text{ rel } X) \to [\Sigma^i(M/X), B/A]$$

with the following properties:

a) $\mathcal{K}_i$ is natural with respect to forgetful homomorphisms and homomorphisms induced by relative $\mathcal{A}$-maps.

b) If $n+i = \dim M+i \geq 6$ and $\mathcal{B} \neq \mathcal{H}$, then $\mathcal{K}_i(\mathcal{B},\mathcal{A};M \text{ rel } X)$ is a bijection.

c) Let $L_k^s(\pi)$ denote the Wall surgery obstruction groups (for simple homotopy equivalences) associated to $\pi = \pi_1(M-X)$, as defined in [37]. Then, if $n+i \geq 6$, there is an exact sequence of groups and homomorphisms

$$\begin{array}{ccc} [\Sigma^{i+1}(M/X), SG/A] \xrightarrow{\theta} L^s_{n+i+1}(\pi) \to & \pi_i(\mathcal{H},\mathcal{A};M \text{ rel } X) & \\ & \downarrow \mathcal{K}_i & \\ & [\Sigma^i(M/X), SG/A] \xrightarrow{\theta} L^s_{n+i}(\pi), & \end{array}$$

where θ is the surgery obstruction homomorphism (cf. [37], p. 10.15).

We prove this theorem in §3.6. In the remainder of this section we state and prove some corollaries.

3.5.6. Corollary: Suppose that $\mathcal{A} \neq \mathcal{D}iff$ and that $\pi_1(M-X) = 0$. Then, under the hypotheses of 3.5.5, the sequence in c) induces the short split-exact sequence

$$0 \to \pi_i(\mathcal{H},\mathcal{A};M \text{ rel } X) \xrightarrow{\mathcal{K}_i} [\Sigma^i(M/X), SG/A] \xrightarrow{\theta} \pi_{n+i}(G/PL) \to 0.$$

$$\underline{\text{Here}},\ \pi_j(G/PL) = \pi_j(SG/SPL) = \begin{cases} 0 \ , & j \text{ odd} \\ \mathbb{Z}_2, & j = 2 (\text{mod } 4) \\ \mathbb{Z} \ , & j = 0 (\text{mod } 4). \end{cases}$$

Proof: That

$$L_j^s(0) = \pi_j(G/PL) \quad (= \pi_j(SG/SPL))$$

is due independently to Sullivan [33] and Wagonner [36]. It remains to show that

$$[\Sigma^k(M/X), SG/A] \xrightarrow{\theta} \pi_{n+k}(SG/SPL),\ k = i, i+1,$$

splits under the stated conditions.

The relative $\mathcal{Q}$-map

$$(M, X) \to (S^n, D_+^n)$$

described in 3.3.4 and 3.3.6 determines a homomorphism

$$\pi_{n+k}(SG/A) \xrightarrow{\lambda} [\Sigma^k(M/X), SG/A].$$

Results of Sullivan [35] and Kirby and Siebenmann [18] imply that $\theta\lambda$ is an isomorphism. Q. E. D.

For certain pairs (M, X), the next corollary reduces the computation of

$$\pi_i(\mathcal{Q}; M \text{ rel } X)$$

to a problem in algebraic topology.

3.5.7. Corollary: Assume the hypotheses of 3.5.6 and suppose that

M-X *admits the structure of a homotopy-associative H-space with homotopy-inverses* (see 3.4.2 and Appendix D). *For every* $i \geq 1$, *define*

$$H_i = [\Sigma^i(M/X), M\text{-}X] \oplus \pi_{n+i}(SG/A)$$
$$K_i = [\Sigma^i(M/X), SG/A].$$

Then, there is an exact sequence

$$\cdots \to H_{i+1} \to K_{i+1} \to \pi_i(\mathcal{A}; M \text{ rel } X) \to H_i \to K_i \to \cdots \to \pi_1(\mathcal{A}; M \text{ rel } X) \to H_1 \to K_1.$$

Proof: We use the following easy algebraic fact: if

$$\cdots \to A_i \xrightarrow{f} B_i \xrightarrow{g} C_i \xrightarrow{\partial} A_{i-1} \xrightarrow{f} B_{i-1} \to \cdots$$

is an exact sequence of abelian groups, and $\{D_i\}$ is an arbitrary sequence of abelian groups, then the sequence

$$\cdots \to A_i \oplus 0 \xrightarrow{f\oplus 0} B_i \oplus D_i \xrightarrow{g\oplus 1} C_i \oplus D_i \xrightarrow{\partial\oplus 0} A_{i-1} \oplus 0 \xrightarrow{f\oplus 0} \cdots$$

is exact.

Let the first sequence be the $(\mathcal{H}; \mathcal{A})$-sequence for (M, X), and let $D_i = \pi_{n+i}(SG/A)$. Then, in the second sequence, replace

$$C_i \oplus D_i = \pi_i(\mathcal{H}; \mathcal{A}; M \text{ rel } X) \oplus \pi_{n+i}(SG/A)$$

by

$$K_i = [\Sigma^i(M/X), SG/A]$$

via the splitting described in the proof of 3.5.6, and replace

$$B_i \oplus D_i = \pi_i(\mathcal{H}; M \text{ rel } X) \oplus D_i$$

by H_i via the isomorphism given by Proposition 3.4.2. Q. E. D.

Remarks: a) It would be interesting to have a direct, natural, homotopy-theoretic definition of the composite

$$[\Sigma^i(M/X), M-X] \subseteq H_i \rightarrow K_i = [\Sigma^i(M/X), SG/A].$$

For example, is it induced by a map $M-X \rightarrow SG/A$?

b) Note that, for (M, X) and i as in 3.5.7, the corollary implies that $\pi_i(\mathcal{A}; M \text{ rel } X)$ is a finitely-generated group which, up to an extension, depends only on the homotopy types of M/X and $M-X$.

3.5.8. Corollary: Suppose that $\pi_1(M-X) = 0$. Then, under the hypotheses of 3.5.5 c) with $\mathcal{A} = \mathcal{D}iff$, we obtain an exact sequence

$$bP_{n+i+1} \rightarrow \pi_i(\mathcal{H}, \mathcal{D}iff; M \text{ rel } X)$$
$$\downarrow K_i$$
$$[\Sigma^i(M/X), SG/SO] \xrightarrow{\theta} \pi_{n+i}(SG/SPL)$$

where $bP_{k+1} \subseteq \Gamma^k$ consists of those classes represented by homotopy spheres that bound π-manifolds.

We omit the argument, which is similar to that for 3.5.6.

3.5.9. Corollary: Let T^n be a PL homotopy n-torus, and suppose that

$n+i \geq 6$. Then:

$$\text{a)}\quad \pi_i(\mathcal{H},\mathcal{PL};T^n) \approx H^{3-i}(T^n;\mathbb{Z}_2)$$

$$\text{b)}\quad \pi_i(\mathcal{PL};T^n) \approx \begin{cases} \mathbb{Z}_2, & i=2 \\ 0, & i>2 \end{cases}$$

Proof: Statement a) can be deduced as in the argument of Wall [38]. Or, we simply remark that a) is essentially a result of Wall [38] and Hsiang and Shaneson [16], together with our proof of 3.5.5 which shows that

$$\pi_i(\mathcal{H},\mathcal{PL};T^n) \approx \mathcal{S}(T^n \times D^i \text{ rel } T^n \times S^{i-1}),$$

where $\mathcal{S}$ is the "structure" group of Sullivan, Wall, Hsiang and Shaneson.

Statement b) follows from a) and Proposition 3.4.3. Q.E.D.

3.5.10. Corollary: $\pi_i(\mathcal{T}op;T^n) = \begin{cases} 0 & , i > 3, n+i \geq 6 \\ 0 \text{ or } \mathbb{Z}_2, & i = 3, n+i \geq 6. \end{cases}$

This follows easily from 3.5.9, the $(\mathcal{T}op,\mathcal{PL})$-sequence for T^n, and the fact, due to Kirby and Siebenmann [18], [20] that STop/SPL is a $K(\mathbb{Z}_2, 3)$.

For some applications, we shall be interested in the homomorphism

$$\pi_i(\mathcal{B},\mathcal{A};S^n \text{rel } D^n_+) \to \pi_i(\mathcal{B},\mathcal{A};M^n \text{ rel } X)$$

induced by the relative $\mathcal{A}$-map

$$(M^n, X) \to (S^n, D^n_+)$$

described in 3.3.6. Although we do not need this fact, we note that, when X is regular, the homomorphism is independent of any unnatural choices that appeared in its construction, because M-X is connected (see 3.3.6).

The Classification Theorem suggests that we can study the above homomorphism by looking at the map

$$\pi_{n+i}(B/A) \to [\Sigma^i(M^n/X), B/A],$$

which is induced by the relative $\mathcal{Q}$-map described in 3.3.6. In fact, when $\mathcal{B} \neq \mathcal{H}$ and $n+i \geq 6$, this second homomorphism is equivalent to the first.

3.5.11. Proposition: Let M^n and X be as in 3.5.5, and suppose that the stable normal bundle of $M^n - X$ is fibre-homotopically trivial. Let the homomorphism

$$\pi_{n+i}(B/A) \xrightarrow{p^*} [\Sigma^i(M^n/X), B/A]$$

be induced by any degree-one map of pairs

$$(M, X) \xrightarrow{p} (S^n, D^n_+).$$

Then, p^* is a split injection.

Proof: Let V be a regularizing manifold for X. By enlarging D^n_+ slightly, if necessary, we may suppose that p is a map of pairs $(M, V) \to (S^n, D^n_+)$. Thus, we have a commutative diagram

$$\begin{array}{ccc} & \xrightarrow{p^*} & [\Sigma^i(M^n/V), B/A] \\ \pi_{n+i}(B/A) & & \downarrow \approx \\ & \xrightarrow[p^*]{} & [\Sigma^i(M^n/X), B/A], \end{array}$$

in which the vertical map is an isomorphism because the collapsing map $M^n/X \to M^n/V$ is a homotopy equivalence. Thus, it suffices to prove that the top homomorphism p^* is a split injection.

Let $M_V = M - \text{int } V$ and imbed $(M_V, \partial M_V)$ in (D^{n+K}, S^{n+K-1}), K large, with tubular neighborhood pair (N, N_0). Let $S = \partial N - \text{int } N_0$. Then, there is a fibre-homotopy equivalence $S \to M_V \times S^{K-1}$ which extends to a homotopy equivalence of pairs

$$(N, \partial N) \xrightarrow{h} (M_V \times D^K, \partial(M_V \times D^K)).$$

We now consider the composite

$$\begin{array}{l} D^{n+K}/S^{n+K-1} \xrightarrow{\pi} N/\partial N \\ \qquad\qquad\qquad\qquad \downarrow \bar{h} \\ \qquad\qquad M_V \times D^K/\partial(M_V \times D^K) \xrightarrow{j} M^n \times D^K/(V \times D^K) \cup (M^n \times S^{K-1}) \\ \qquad\qquad\qquad\qquad\qquad\qquad\qquad\qquad \downarrow \bar{p} \\ \qquad\qquad\qquad\qquad\qquad\qquad S^n \times D^K/(D^n_+ \times D^K) \cup (S^n \times S^{K-1}). \end{array}$$

Here π is the standard Thom collapsing map, $\bar{h}$ and $\bar{p}$ are induced by h and p respectively, and j is induced by the inclusion $M_V \times D^K \subseteq M^n \times D^K$.

Thus, for sufficiently large K, we obtain maps

$$S^{n+K} \xrightarrow{\bar{q}} \Sigma^K(M/V) \xrightarrow{\bar{p}} S^{n+K}$$

such that $\overline{pq}$ has degree one. It follows that for any pointed space Y, the homomorphism

$$\pi_{n+K}(Y) \xrightarrow{p^*} [\Sigma^K(M/V), Y]$$

is split by q^*. Given i, choose K larger than i. Then, we have that for any pointed space Y,

$$\pi_{n+i}(\Omega^{K-i}Y) \xrightarrow{p^*} [\Sigma^i(M/V), \Omega^{K-i}Y]$$

is split by q^*.

It now remains only to observe that, according to results of Boardman and Vaught [2], each space B/A is an infinite loop space. That is, for any value of K-i, there is a pointed space Y with $B/A = \Omega^{K-i}Y$. Q. E. D.

3.5.12. Corollary: *Let* M^n *and* X *be as in* 3.5.5 *such that* M^n-X *admits the structure of an* H-*space. Let* p^* *be as in* 3.5.11. *Then* p^* *is a split injection.*

Proof: $M^n - X$ has fibre-homotopically trivial stable normal bundle. The proof parallels its Lie group analogue. Q. E. D.

See Appendix D for examples of pairs (M^n, X) satisfying (more than) the hypotheses of 3.5.12.

By combining the Exactness Theorem with Proposition 3.4.1, we see that $\pi_i(\mathcal{B}, \mathcal{A}; S^n \text{rel } D^n_+)$ is trivial unless $\mathcal{A} = \mathcal{D}iff$, and then $\pi_i(\mathcal{B}, \mathcal{D}iff; S^n \text{rel } D^n_+) \approx \Gamma^{n+i}$.

3.5.13. Corollary: *Let* M^n *and* X *be as in* 3.5.11, *and let* $p : (M^n, X) \to (S^n, D^n_+)$ *be a relative* $\mathcal{D}iff$-*map. Suppose that* $n+i \geq 6$.

Then:

a) If $\mathcal{B} \neq \mathcal{H}$, then

$$\Gamma^{n+i} \approx \pi_i(\mathcal{B}, \mathcal{D}iff ; S^n \text{rel } D^n_+) \xrightarrow{p^\#} \pi_i(\mathcal{B}, \mathcal{D}iff ; M^n \text{rel } X)$$

is a split injection.

b) The kernel of

$$\Gamma^{n+i} \approx \pi_i(\mathcal{H}, \mathcal{D}iff ; S^n \text{rel } D^n_+) \xrightarrow{p^\#} \pi_i(\mathcal{H}, \mathcal{D}iff ; M \text{ rel } X)$$

is contained in bP_{n+i+1} (cf. 3.5.8).

Statement a) follows immediately from 3.5.5 b) and 3.5.11. Statement b) follows from 3.5.5 a), 3.5.8, and 3.5.11: for 3.5.5 a) and 3.5.11 imply that kernel $p^\# \subseteq$ kernel $(K_i : \pi_i(\mathcal{H}, \mathcal{D}iff ; S^n \text{rel } D^n_+) \to \pi_{n+i}(SG/SO))$, and 3.5.8 implies that this latter kernel lies in the image of bP_{n+i+1} (which, in this case is injected). Q.E.D.

We conclude this section with a remark concerning the dimension restriction in the Classification Theorem. The condition $n+i \geq 6$ is necessary, both for our proof, in which it permits us to appeal to the s-cobordism theorem and to apply results of Sullivan, Wall, Kirby, and Siebenmann that have this restriction, and because there is a simple counterexample:

$$\pi_2(\mathcal{T}op, \mathcal{PL} ; S^1 \text{rel } D^1_+) = 0 \qquad \text{(cf. 3.4.1).}$$

$$\pi_3(TOP/PL) = \mathbb{Z}_2 \qquad \text{(see [20]).}$$

§3.6 PROOF OF THE CLASSIFICATION THEOREM

3.6.1. General outline of proof

We shall define maps $\mathcal{K}_i$ and σ_i that fit into the commutative diagram

$$\begin{array}{ccc} \pi_i(\mathcal{B},\mathcal{A};M \text{ rel } X) & \xrightarrow{\mathcal{K}_i} & \\ \sigma_i \downarrow & & [\Sigma^i(M/X), B/A] \\ \mathcal{S}_i(\mathcal{B},\mathcal{A};M \text{ rel } X) & \xrightarrow{\eta_i} & \end{array}$$

Here, $\mathcal{S}_i(\mathcal{B},\mathcal{A};M \text{ rel } X)$ is the structure group of Sullivan [33], Wall [37], and others (e.g., [24]), and η_i is the classification map defined by these authors.

Actually, we define $\mathcal{K}_i$ (3.6.6) and σ_i (3.6.9) on the level of representatives, where commutativity of the corresponding diagram is immediate from definitions. We show that $\mathcal{K}_i$, if well-defined, is a homomorphism with the naturality properties required by part a) of the Classification Theorem (3.6.6, 3.6.7). Then we show that σ_i is a well-defined injection which is bijective when $n+i \geq 6$. The various remaining facts about $\mathcal{K}_i$ stated in the Classification Theorem--including the fact that it is well-defined--then follow from the corresponding facts about η_i due to the above-mentioned authors. We elaborate on this in 3.6.15.

3.6.2. Reduction to a special case

Lemma: Suppose that the Classification Theorem holds for all X that are compact, codimension-0, locally-flat, $\mathcal{A}$-submanifolds of M. Then the theorem holds for all regular X.

Proof: Let X be regular, and let V be a regularizing manifold for X. By hypothesis, the Classification Theorem holds for all such V. We define $\mathcal{K}_i(\mathcal{B},\mathcal{A};M \text{ rel } X)$ by the commutative diagram

$$\begin{array}{ccc} \pi_i(\mathcal{B},\mathcal{A};M \text{ rel } X) & \xrightarrow{\mathcal{K}_i} & [\Sigma^i(M/X), B/A] \\ \approx\uparrow & & \approx\uparrow \\ \pi_i(\mathcal{B},\mathcal{A};M \text{ rel } V) & \xrightarrow{\mathcal{K}_i} & [\Sigma^i(M/V), B/A] \end{array} .$$

The vertical isomorphism on the left comes from 3.5.3; that on the right is due to the fact that the collapse

$$M/X \to M/V$$

is a homotopy equivalence.

Parts b) and c) of the Classification Theorem follow immediately, as does naturality with respect to change of categories:

$$(\mathcal{B},\mathcal{A}) \to (\overline{\mathcal{B}},\overline{\mathcal{A}}), \text{ when } (\mathcal{B},\mathcal{A}) \leq (\overline{\mathcal{B}},\overline{\mathcal{A}}).$$

It remains to prove naturality with respect to homomorphisms induced by relative $\mathcal{A}$-maps

$$f : (M_1, X_1) \to (M_2, X_2).$$

Let V_2 be a regularizing manifold for X_2. It is easy to see that a regularizing manifold V_1 for X_1 may be chosen, such that $V_1 \subset f^{-1}(V_2)$, and one readily checks that

$$f : (M_1, V_1) \to (M_2, V_2)$$

is a relative $\mathcal{A}$-map.

The desired result now follows, provided that we can show that the definition of $\mathcal{K}_i(\mathcal{B}, \mathcal{A}; M \text{ rel } X)$ does not depend on the choice of regularizing manifold V. But, these manifolds are ordered by inclusion, and, for any two such, V_1 and V_2, there is a third,

$$V_3 \subset \text{int}\,(V_1 \cap V_2).$$

From this and the naturality part of the theorem applied to the V's, we deduce the independence of $\mathcal{K}_i(\mathcal{B}, \mathcal{A}; M \text{ rel } X)$. Q. E. D.

From now on in this section, X will be a compact, co-dimension-0, locally-flat $\mathcal{A}$-submanifold of M, with M-X connected.

3.6.3. Differentials and stable differentials in $\mathcal{D}iff$, $\mathcal{PL}$, and $\mathcal{T}op$

In each category $\mathcal{A} = \mathcal{D}iff$, $\mathcal{PL}$, or $\mathcal{T}op$ there are defined notions of (oriented) microbundle and map of microbundles. For detailed definitions and important properties of these objects, we refer the reader to [27], [10], [24], [22]. Here, we present only some notation and some facts that we need.

The most important example of an $\mathcal{A}$-microbundle is the tangent microbundle τ_M of the $\mathcal{A}$-manifold M (see [27]), and the most important example of a map of $\mathcal{A}$-microbundles is the differential

$$dh : \tau_{M_1} \to \tau_{M_2}$$

of a map $h : M_1 \to M_2$ in $\mathcal{A}$. Of course, when $\mathcal{A} = \mathcal{D}iff$, there are classical notions of tangent bundle and differential. These notions are equivalent to their microbundle counterparts (see [17], Chapter II), however, and so we shall use the latter to keep our treatment uniform.

Two other important examples of $\mathcal{A}$-microbundles are: (1) the normal microbundle ν_M of an $\mathcal{A}$-manifold M that is locally-flatly $\mathcal{A}$-imbedded in some high-dimensional Euclidean space; (2) the trivial microbundle $\varepsilon_n(X)$ over X with fibre $\mathbb{R}^n$, $n \geq 0$. Results of Milnor [27], Lashof and Rothenberg [24], §5, Haefliger and Wall [9], and Hirsch [10] show that the classical existence and uniqueness results for smooth tubular neighborhoods carry over, mutatis mutandi, to ν_M, provided that the receiving Euclidean space has sufficiently high dimension. We shall not concern outselves with the question of exactly how high this dimension must be (but it must be at least as high as 2dim M+2, in general).

If the fibre-dimension of a microbundle is high enough, in comparison with the dimension of its base space, then its equivalence class is completely determined by its stable equivalence class. We call such a microbundle stable. Clearly, any microbundle can be stabilized: that is, it can be transformed into a stable microbundle by adding (i.e., forming the Whitney sum with) a trivial bundle ε_n, n sufficiently large. Again we are not interested in how large n must be. Similar comments apply to maps of $\mathcal{A}$-microbundles. It will be convenient to preserve the notation τ_M and dh for a stabilized tangent bundle and differential and ν_M for the stable normal bundle.

Let $\tau_i = \tau_{M_i}$ and $\nu_i = \nu_{M_i}$, $i = 1, 2$, and note that dh factors:

$$\begin{array}{ccc} \tau_1 & \xrightarrow{dh} & \tau_2 \\ {\scriptstyle \tau h} \searrow & & \nearrow {\scriptstyle \bar{h}} \\ & h^* \tau_2 & \end{array}$$

Here $\bar{h}$ is the standard microbundle map covering h that is provided by the pullback construction.

There is a well-known procedure for passing from stable $\mathcal{A}$-microbundle equivalences

$$\tau_1 \to h^* \tau_2$$

to stable $\mathcal{A}$-microbundle equivalences

$$\nu_1 \to h^* \nu_2.$$

This procedure induces a 1-1 correspondence on the level of stable $\mathcal{A}$-microbundle homotopy classes of stable bundle equivalences (cf. Browder [4], Chapter I.4). We sketch the method.

There exist trivializations

$$\varepsilon_K(M_1) \xrightarrow{t_1} \nu_1 + \tau_1$$

$$\varepsilon_K(M_2) \xrightarrow{t_2} \nu_2 + \tau_2,$$

where K is some large natural number. We can canonically identify $\varepsilon_K = \varepsilon_K(M_1)$ with the pullback $h^*\varepsilon_K(M_2)$, so that the pullback h^*t_2 is a map

$$\varepsilon_K \to h^*\nu_2 + h^*\tau_2.$$

Now, given any equivalence $f : \tau_1 \to h^*\tau_2$, we form the composite

$$(1)\qquad \begin{array}{ccc} \varepsilon_K + h^*\nu_2 & \xrightarrow{t_1+1} & \nu_1 + \tau_1 + h^*\nu_2 \\ & & \downarrow 1+f+1 \\ & & \nu_1 + h^*\tau_2 + h^*\nu_2 \xrightarrow{1+h^*t_2^{-1}} \nu_1 + \varepsilon_K, \end{array}$$

where we abuse notation slightly by not preceding $1 + h^*t_2^{-1}$ with an appropriate switch of summands. The microbundle map corresponding to f, then, is the inverse of this composite; call it g. If we now apply the same procedure to g, using the trivializations t_1 and t_2 again, the result will be stably $\mathcal{A}$-microbundle homotopic to f. We shall not need these facts, and so, we do not give details.

We denote by

$$\nu h : \nu_1 \to h^*\nu_2$$

the equivalence corresponding to τh under the above construction, and we let nh be the composite

$$\nu_1 \xrightarrow{\nu h} h^*\nu_2 \xrightarrow{\overline{h}} \nu_2$$

covering h. Note that whereas τh and dh depend only on h, νh <u>depends on the choice of</u> t_1 <u>and</u> t_2.

We now refer to diagram (1), obtaining the following commutative diagram

$$\begin{array}{ccc}
\tau_1 + \nu_1 + \varepsilon_K & \xrightarrow{f + g} & h^*\tau_2 + \varepsilon_K + h^*\nu_2 \\
\Big\downarrow {\scriptstyle 1 + 1 + h^*t_2} & & \Big\downarrow {\scriptstyle 1 + t_1 + 1} \\
\tau_1 + \nu_1 + h^*\tau_2 + h^*\nu_2 & \xleftarrow{f^{-1} + 1 + f + 1} & h^*\tau_2 + \nu_1 + \tau_1 + h^*\nu_2 \\
{\scriptstyle f + 1 + 1 + 1} \searrow & & \swarrow {\scriptstyle 1 + 1 + f + 1} \\
& h^*\tau_2 + \nu_1 + h^*\tau_2 + h^*\nu_2 & .
\end{array}$$

It follows easily, after switching appropriate summands--and this only alters the diagram by $\mathcal{A}$-microbundle homotopy--and setting $f = \tau h$, $g = \nu h$, that the following diagram of stable equivalences commutes, at least up to $\mathcal{A}$-microbundle homotopy:

$$\begin{array}{ccc}
\tau_1 + \nu_1 & \xrightarrow{\tau h + \nu h} & h^*\tau_2 + h^*\nu_2 \\
{\scriptstyle t_1} \nwarrow & & \nearrow {\scriptstyle h^*t_2} \\
& \varepsilon_K & .
\end{array} \qquad (2)$$

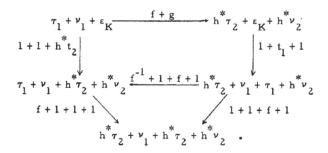

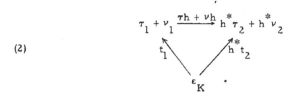

Observe that differentials are natural with respect to change of categories by a forgetful functor. In fact, there is a relative naturality: namely, if $h : (M_1, V_1) \to (M_2, V_2)$ is a map of pairs in $\mathcal{B}$ such that $h(V_1) = V_2$, where V_i is a locally-flat, codimension-0, $\mathcal{A}$-submanifold of M_i, $i = 1, 2$, with $h|U : U \to h(U)$ in $\mathcal{Q}$, for some neighborhood U of V_1, then, $dh|V_1$ is the $\mathcal{Q}$-differential $d(h|V_1)$.

3.6.4. Stable differentials of maps in $\mathcal{H}$

In this section, $\mathcal{Q} = \mathcal{D}iff$, $\mathcal{PL}$, or $\mathcal{T}op$, M_1 and M_2 are $\mathcal{Q}$-manifolds, and $h : M_1 \to M_2$ is a map in $\mathcal{H}$. The definitions of dh and τh are more complicated here, requiring results of Spivak [32] and Browder [4] on stable spherical normal fibre spaces. We proceed in precisely the reverse of the order of 3.6.3, first defining νh and nh, and then obtaining τh and dh from these via the procedure described in diagram (1) of 3.6.3. We now give some details.

We restrict ourselves to compact $\mathcal{Q}$-manifolds M_i^n with stable normal microbundles ν_i, $i = 1, 2$, such that the fibre-dimension k of ν_i is large relative to n. Let $K = k + n$. We may always choose ν_i to be an $\mathcal{Q}$-bundle: when $\mathcal{Q} = \mathcal{D}iff$, we use the usual normal vector bundle; when $\mathcal{Q} = \mathcal{PL}$, we use the PL version of Kister's theorem, due to Kuiper and Lashof [22]; and, when $\mathcal{Q} = \mathcal{T}op$, we use Kister's theorem [21] together with the fact that each $\mathcal{T}op$ manifold is an ANR.*

---

*Kister's theorem requires the base space to be a polyhedron. Imbed the given $\mathcal{T}op$ manifold M in some Euclidean space, and apply Kister's theorem to the pullback of ν_M under a neighborhood retraction. Restrict the result to M. This is the new ν_M.

Thus, we may talk about the pair of total spaces $(E_i, \overline{E}_i)$ of the bundle-pair $(\nu_i, \nu_i/\partial M_i)$.

Henceforth, we suppose that $(M_i, \partial M_i) \subseteq (D^K, S^{K-1})$, as locally-flat $\mathcal{Q}$-submanifolds. Provided that k (and, hence, K) is suitably large, this is always possible and will not restrict generality. We also identify M_i with the zero-section of ν_i. Choose locally-flat $\mathcal{Q}$-imbeddings

$$j_i : (E_i, \overline{E}_i) \to (D^K, S^{K-1}), \qquad i = 1, 2,$$

subject only to the condition

$$j_i(x) = x, \qquad \text{for all } x \in M_i.$$

These imbeddings determine classes

$$\theta_i \in \pi_K(T(\nu_i), T(\nu_i/\partial M_i)), \qquad i = 1, 2,$$

by the usual Thom construction. We now use the $\mathcal{H}$-map $h : M_1 \to M_2$ and the corresponding bundle map

$$\overline{h} : h^* \nu_2 \to \nu_2$$

to pull θ_2 back to a class

$$h^* \theta_2 \in \pi_K(T(h^* \nu_2), T(h^* \nu_2/\partial M_1)).$$

According to Spivak's Uniqueness Theorem, there is defined a stable fibre-homotopy equivalence*

$$\nu_1 \xrightarrow{\nu h} h^* \nu_2$$

such that $T(\nu h)_*(\theta_1) = h^*\theta_2$ and such that this last equation characterizes νh up to stable fibre homotopy (see [32] or [4]).

Now, the imbeddings j_i also determine $\mathfrak{a}$-trivializations

$$\varepsilon_K \xrightarrow{t_i} \nu_1 + \tau_1.$$

Roughly speaking, t_i^{-1} is just $dj_i : \tau_{E_i}/M_i \to \tau_{D^K}/M$. Here, we use the standard equivalence,

$$\tau_{E_i}/M_i \approx \nu_1 + \tau_1.$$

The construction of 3.6.3 now produces a stable fibre-homotopy

*A proper map f is one for which f^{-1} (compact) is compact. A stable fibre-homotopy equivalence of Euclidean bundles $\xi \to \eta$ (over a given base space) is a map $f : E(\xi + \varepsilon_n) \to E(\eta + \varepsilon_n)$, for some n, for which there exists a homotopy-inverse, all maps and homotopies being proper and fixing fibres setwise. Stable fibre-homotopy is defined analogously. (A similar definition can be formulated for microbundles.) This definition differs from the standard one (e.g., see [32]), in which one passes to the associated stable sphere-bundles, but the two are equivalent (see Lashof and Rothenberg [25], p. 82).

equivalence

$$\tau_1 \xrightarrow{\tau h} h^* \tau_2 .$$

The fibre maps nh and dh are related to νh and τh as in 3.6.3.

Different choices of j_i will determine different θ_i, in general, and also different νh. But, they also determine different trivializations t_i. These differences "cancel" in the construction of τh, so that τh depends, up to stable fibre-homotopy, on h alone. We do not need this fact here, and since the proof is cumbersome, we omit it.

The differential dh and the fibre-homotopy equivalence τh are natural with respect to change of categories. In particular, the relative naturality described in 3.6.3 now also applies when $\mathcal{B} = \mathcal{H}$. There is a proof required in this case, however. The proof consists of two observations:

(i) The above construction of dh and τh is natural with respect to inclusions $V_i \subseteq M_i$, V_i as in 3.6.3. This implies that $dh|V_1$ is the $\mathcal{H}$-differential of $h|V_1$.

(ii) The $\mathcal{A}$-differential of $h|V_1$ satisfies the hypotheses of Spivak's Uniqueness Theorem. Thus, by the Uniqueness Theorem, $dh|V_1$ is stably fibre-homotopic to the $\mathcal{A}$-differential of $h|V_1$.

3.6.5. The map $(\mathcal{B},\mathcal{A})_i(M \text{ rel } X) \xrightarrow{k_i} [\Sigma^i(M/X), B/A]$

Let τ_M and ν_M denote the stable tangent and normal microbundles, respectively, of $M \times \mathbb{R}^i_+$. Fix a stable $\mathcal{A}$-trivialization

$$\epsilon \xrightarrow{t} \tau_M + \nu_M .$$

Given $f \in (\mathcal{B},\mathcal{A})_i(M \text{ rel } X)$, we define the stable $\mathcal{B}$-trivialization t_f to be the composite

$$\epsilon \xrightarrow{t} \tau_M + \nu_M \xrightarrow{\tau f + 1} f^* \tau_M + \nu_M, \quad (f^* = (f \times id)^*).$$

(Although $M \times \mathbb{R}^i_+$ is not compact, supp f is, so that the description of τf given in 3.6.4 is easily adapted to this case.) It will be convenient to let $\mathbb{R}^i_+ \cup \infty$ denote the one-point compactification of $\mathbb{R}^i_+$.

The salient features of

$$\epsilon \xrightarrow{t' = t_f} \alpha = f^* \tau_M + \nu_M$$

are these:

a) α <u>is an $\mathcal{A}$-microbundle that coincides with</u> $\tau_M + \nu_M$ <u>over a</u> neighborhood of $(M \times \infty) \cup X \times \mathbb{R}^i_+$;

b) t' <u>is a $\mathcal{B}$-trivialization of</u> α <u>that is an $\mathcal{A}$-trivialization on a neighborhood of</u> $(M \times \mathbb{R}^{i-1}) \cup (X \times \mathbb{R}^i_+)$.

There is a standard definition of $(\mathcal{B},\mathcal{A})$-concordance between any two trivializations t' that satisfy a) and b) (e.g., see [12], [16], [24], [33], [37]) and the concordance classes are in natural 1-1 correspondence with

$$[\Sigma^i(M/X), B/A].$$

Let $\chi(t')$ be the homotopy class corresponding to t'. Then, k_i is defined by

$$k_i(f) = \chi(t_f).$$

3.6.6. The homomorphism $\pi_i(\mathcal{B},\mathcal{A};M \text{ rel } X) \xrightarrow{\mathcal{K}_i} [\Sigma^i(M/X), B/A]$.

It is not hard to check directly that k_i well-defines a map

$$\pi_i(\mathcal{B},\mathcal{A};M \text{ rel } X) \xrightarrow{\mathcal{K}_i} [\Sigma^i(M/X), B/A].$$

Since, however, we prefer not to specify in detail the relation of concordance between trivializations mentioned above in 3.6.5, we omit the verification. In any case, the well-definition follows from results cited in later paragraphs.

Using juxtaposition (see 3.2.6), it is easy to see that $\mathcal{K}_i$ defines a homomorphism when $i \geq 2$. When $i = 1$, we can show that composition is taken by $\mathcal{K}_i$ to Whitney sum. We omit the proof, noting that the result has been asserted in [34] and [15]. (See [24], §5, Lemma, for a related proof.)

3.6.7. Naturality.

The naturality of $\mathcal{K}_i$ with respect to change-of-categories follows easily from definitions. Here, we demonstrate naturality with respect to homomorphisms induced by relative $\mathcal{A}$-maps (cf. Definition 3.3.4).

Let

$$h : (M_1, X_1) \to (M_2, X_2)$$

be such a map, and let $U \subseteq M_1 - X_1$ be the $\mathcal{Q}$-isomorphic image of $M_2 - X_2$ under $h^{-1}/M_2 - X_2$ (cf. Definition 3.3.4). Let $H = h \times id_{\mathbb{R}^i_+}$. For any $f \in (\mathcal{B}, \mathcal{Q})_i(M_2 \text{ rel } X_2)$, we compare the $\mathcal{B}$-trivializations

$$\epsilon_{(1)} \xrightarrow{t_{h^\# f}} (h^\# f)^* \tau_{M_1} + \nu_{M_1}$$

and

$$\epsilon_{(1)} = H^* \epsilon_{(2)} \xrightarrow{H^* t_f} H^* f^* \tau_{M_2} + H^* \nu_{M_2}.$$

Here, $\epsilon_{(\alpha)}$ is the standard trivial bundle over $M_\alpha \times \mathbb{R}^i_+$, $\alpha = 1, 2$, of high fibre-dimension; there is a canonical identification of $\epsilon_{(1)}$ and $H^* \epsilon_{(2)}$. $H^* t_f$ is the pullback of t_f.

Let $t_\alpha : \epsilon_{(\alpha)} \to \tau_{M_\alpha} + \nu_{M_\alpha}$, $\alpha = 1, 2$, be the fixed $\mathcal{Q}$-trivializations used to define k_i in 3.6.5.* Use these trivializations in the manner described in 3.6.3 to define the $\mathcal{Q}$-isomorphism of stable $\mathcal{Q}$-microbundles

$$\nu_{M_1} | U \times \mathbb{R}^i_+ \xrightarrow{\nu H} H^* \nu_{M_2} | U \times \mathbb{R}^i_+.$$

Then, the following diagram commutes, up to $\mathcal{Q}$-microbundle homotopy (see diagram (2) of 3.6.3):

(1)
$$\begin{array}{ccc} & \xrightarrow{t_1} & \tau_{M_1} + \nu_{M_1} \\ \epsilon_{(1)} & & \downarrow \tau H + \nu H \\ & \xrightarrow{H^* t_2} & H^* \tau_{M_2} + H^* \nu_{M_2}, \end{array}$$

with everything restricted to the open set $U \times \mathbb{R}^i_+$.

Now, a direct check shows that the following diagram commutes (over $U \times \mathbb{R}^i_+$):

$$
(2) \qquad
\begin{array}{ccc}
\tau_{M_1} + \nu_{M_1} & \xrightarrow{\tau(h^{\#} f)+1} & (h^{\#} f)^* \tau_{M_1} + \nu_{M_1} \\
\Big\downarrow \tau H + \nu H & & \Big\downarrow (h^{\#} f)^* \tau H + \nu H \\
H^* \tau_{M_2} + H^* \nu_{M_2} & \xrightarrow{H^* \tau(f)+1} & H^* f^* \tau_{M_2} + H^* \nu_{M_1}
\end{array}
$$

Here, we use the relations

$$\tau(ab) = (b^* \tau(a)) \circ \tau(b)$$

and

$$h^{\#} f \,|\, U \times \mathbb{R}^i_+ = H^{-1} f H \,|\, U \times \mathbb{R}^i_+ .$$

Combining (1) and (2) we obtain a diagram

$$
(3) \qquad
\begin{array}{ccc}
 & \overset{t_{h^{\#} f}}{\nearrow} & (h^{\#} f)^* \tau_{M_1} + \nu_{M_1} \\
\epsilon & (1) & \Big\downarrow G \\
 & \underset{H^* t_f}{\searrow} & H^* f^* \tau_{M_2} + H^* \nu_{M_2}
\end{array} ,
$$

defined and commutative (up to $\mathcal{B}$-microbundle homotopy) over $U \times \mathbb{R}^i_+$, where G is a certain $\mathcal{A}$-isomorphism. Note that over the complement of the compact set

$$\operatorname{supp} h^{\#} f \subset U \times \mathbb{R}^i_+ ,$$

diagram (3) reduces to diagram (1), which consists entirely of

$\mathcal{A}$-microbundle maps, and the $\mathcal{B}$-microbundle homotopy-commutativity restricts to $\mathcal{A}$-microbundle homotopy-commutativity. Similar reductions from $\mathcal{B}$ to $\mathcal{A}$ apply to restrictions over $U \times \mathbb{R}^{i-1}$.

To extend G over all of $M_1 \times \mathbb{R}^i_+$, choose a compact, codimension-0, locally-flat $\mathcal{A}$-submanifold $W \subset U \times \mathbb{R}^i_+$, such that, setting $V =$ closure $(\partial W - (U \times \mathbb{R}^{i-1}))$, there is a relative collar pair $(C, C \cap (U \times \mathbb{R}^{i-1}))$ for $(V, \partial V)$ in $(U \times \mathbb{R}^i_+, U \times \mathbb{R}^{i-1})$, with $\text{supp } h^{\#} f \subset W - C$. The existence of such a W can be deduced, for example, from the fact that U is $\mathcal{A}$-isomorphic to the interior of the manifold $M_2 - \text{int } X_2$. We shall actually extend $G/W - \text{int } C$ over $M \times \mathbb{R}^i_+$.

Over the complement of W, define G so that (3) actually commutes. Fill in the strip over C by the given $\mathcal{A}$-microbundle homotopy between

$$\tau H + \nu H | \partial W \quad \text{and} \quad (H^* t_2) \circ t_1^{-1} | \partial W.$$

We leave to the reader the check that G, so defined, makes (3) commute up to $\mathcal{B}$-microbundle homotopy, reducing to $\mathcal{A}$ near $(M \times \mathbb{R}^{i-1}) \cup (X_1 \times \mathbb{R}^i_+)$. This shows that

$$H^* t_f \quad \text{and} \quad t_{h^{\#} f}$$

are concordant, thus proving naturality.

3.6.8. Definition of $\mathcal{S}_i(\mathcal{B},\mathcal{Q};M \text{ rel } X)$.

The following definitions are patterned after those of Sullivan [33] and Wall [37], but they have their origin in the work of Lashof and Rothenberg [24] and Mazur and Hirsch [12]. See [16] for a good short exposition in the $(\mathcal{H};\mathcal{PL})$ case. Let V be a compact manifold in $\mathcal{Q}$. Recall that M has the same meaning as in 3.5.5 and that X is a compact, codimension-0, locally-flat $\mathcal{Q}$-submanifold of M (cf. 3.6.2).

a) A $(\mathcal{B},\mathcal{Q})$-structure for $(V,\partial V)$ is a map of pairs

$$(V,\partial V) \xrightarrow{f} (N,\partial N)$$

such that

(i) N is a compact manifold in $\mathcal{Q}$,

(ii) $f|\partial V$ is a map in $\mathcal{Q}$,

(iii) f is a map in $\mathcal{B}$.

When $\mathcal{B} = \mathcal{H}$, we require additionally that f be a simple homotopy equivalence.

Two $(\mathcal{B},\mathcal{Q})$-structures f_0, f_1 for $(V,\partial V)$ with ranges N_0, N_1 are $\mathcal{S}$-equivalent if there exists a map of pairs $h : (N_0, \partial N_0) \to (N_1, \partial N_1)$ in $\mathcal{Q}$ such that hf_0 is $\mathcal{S}$-concordant rel boundary to f_1. This means that there is a map of pairs in $\mathcal{B}$

$$(V,\partial V) \times [0,1] \xrightarrow{F} (N_1, \partial N_1) \times [0,1]$$

such that

(iv) $F|\partial V \times [0,1] = (f_1|\partial V) \times id_{[0,1]}$

(v) $F|V \times 0 = hf_0 \times 0$

(vi) $F|V \times 1 = f_1 \times 1$.

$\mathcal{S}$-equivalence is easily shown to be an equivalence relation on the $(\mathcal{B},\mathcal{Q})$-structures for $(V, \partial V)$. Let $\mathcal{S}(\mathcal{B},\mathcal{Q};V, \partial V)$ be the set of equivalence classes.

Remark 1: In Sullivan's formulation [33], the arrows go the other way. To provide a uniform treatment, we follow the convention of Lashof and Rothenberg [24], which is, to some extent, forced by the peculiarities of the $(\mathcal{PL}, \mathcal{D}iff)$ case (see Appendix C). The same holds for our conventions in 3.6.3 - 3.6.7.

Remark 2: Endow ∂V with an arbitrary but fixed $\mathcal{Q}$-collaring in V. We consider collared $(\mathcal{B},\mathcal{Q})$-structures on $(V, \partial V)$, which are $(\mathcal{B},\mathcal{Q})$-structures

$$(V, \partial V) \xrightarrow{f} (N, \partial N)$$

that are stationary near ∂V with respect to the preferred collar of ∂V and a preferred $\mathcal{Q}$-collaring of $\partial N \subseteq N$ (which is part of the structure).* We can similarly define the notions of collared $\mathcal{S}$-equivalence, which, again is an equivalence relation, and denote the set of equivalence classes by $\mathcal{S}_{col}(\mathcal{B},\mathcal{Q};V, \partial V)$. Then, using an argument similar to that of 3.3.7,

---

*See 3.1.3.

we can show that the forgetful map

$$\mathcal{S}_{col}(\mathcal{B},\mathcal{A};V,\partial V) \to \mathcal{S}(\mathcal{B},\mathcal{A};V,\partial V)$$

is a bijection. We make use of this fact later.

b) Let D^i be the standard unit ball in $\mathbb{R}^i$. Then $V \times D^i$ is again a compact manifold in $\mathcal{A}$. Define

$$\mathcal{S}_i(\mathcal{B},\mathcal{A};V,\partial V) \equiv \mathcal{S}(\mathcal{B},\mathcal{A};V \times D^i, \partial(V \times D^i)).$$

c) Finally, let M_X = M-interior X, and define

$$\mathcal{S}_i(\mathcal{B},\mathcal{A};M \text{ rel } X) \equiv \mathcal{S}_i(\mathcal{B},\mathcal{A};M_X, \partial M_X).$$

d) When $i \geq 2$, $\mathcal{S}_i(\mathcal{B},\mathcal{A};M \text{ rel } X)$ can be endowed with a natural group structure by pasting together $(\mathcal{B},\mathcal{A})$-structures. We shall not define this (directly), however, Sullivan [33] defines a group structure on $\mathcal{S}(\mathcal{B},\mathcal{A};V,\partial V)$ when $(\mathcal{B},\mathcal{A}) = (\mathcal{H},\mathcal{D}iff)$ or $(\mathcal{H},\mathcal{PL})$ and V and ∂V are 1-connected.

Since the pair $(\mathcal{B},\mathcal{A})$ will be fixed throughout this section, we shall henceforth suppress it from the notation whenever we can. For example, $\mathcal{S}_i(\mathcal{B},\mathcal{A};M \text{ rel } X)$ becomes $\mathcal{S}_i(M \text{ rel } X)$. "$(\mathcal{B},\mathcal{A})$-structures" becomes "structures".

3.6.9. The function $\sigma : \{(\mathcal{B},\mathcal{A})\text{-maps}\} \to \{\text{structures}\}$

Choose an arbitrary but fixed point

$$* \in \text{int } D_+^{i-1} \subsetneq S^{i-1} \subsetneq D^i$$

and an arbitrary but fixed map

$$\lambda : (\mathbb{R}^i_+, \mathbb{R}^{i-1}) \to (D^i, S^{i-1})$$

which is an $\mathcal{A}$-isomorphism onto $D^i - \{*\}$.

Given $f \in (\mathcal{B}, \mathcal{A})_i (M \text{ rel } X)$, let

$$f_X \text{ be } f\big|M_X \times \mathbb{R}^i_+ = f\big|(M-\text{int } X) \times \mathbb{R}^i_+,$$

viewed as a self map of $M_X \times \mathbb{R}^i_+$. Then, let

$$\sigma(f) : (M_X \times D^i, \partial(M_X \times D^i)) \to (M_X \times D^i, \partial(M_X \times D^i))$$

be the $(\mathcal{B}, \mathcal{A})$-structure determined by compactifying

$$(\text{id}_{M_X} \times \lambda) \circ f_X \circ (\text{id}_{M_X} \times \lambda^{-1}).$$

The association $f \to \sigma(f)$ is a function

$$(\mathcal{B}, \mathcal{A})_i (M \text{ rel } X) \overset{\sigma}{\to} \{\text{structures on } (M_X \times D^i, \partial(M_X \times D^i)\}$$

which satisfies $\sigma(fg) = \sigma(f) \circ \sigma(g)$.[*] Clearly, image σ consists of all structures f such that

(i)   range $f = (M_X \times D^i, \partial(M_X \times D^i))$

(ii) $(M_X \times \{*\}) \cup (\partial M_X \times D^i)$ does not meet supp f.

Note that if $f \in (\mathcal{B}, \mathcal{A})_i (M \text{ rel } X)$ is stationary near $M \times \mathbb{R}^{i-1} \subseteq M \times \mathbb{R}^i_+$, then $\sigma(f)$ is a collared $(\mathcal{B}, \mathcal{A})$-structure on $(M_X \times D^i, \partial(M_X \times D^i))$.

---

[*] That $\sigma(f)$ is a simple homotopy equivalence when $\mathcal{B} = \mathcal{H}$ is due to the fact that $\sigma(f)\big|M_X \times \{*\}$ is the identity map.

3.6.10. Lemma: *If* $\sigma(f_0)$ *and* $\sigma(f_1)$ *are* $\mathcal{S}$-*equivalent, then* f_0 *and* f_1 *are* $(\mathcal{B},\mathcal{A})$-*concordant* rel X.

3.6.11. Lemma: *If* f_0 *and* f_1 *are* $(\mathcal{B},\mathcal{A})$-*concordant* rel X, *then* $\sigma(f_0)$ *and* $\sigma(f_1)$ *are* $\mathcal{S}$-*equivalent.*

3.6.12. Lemma: *Suppose that* $n+i \geq 6$, $i \geq 1$. *Then every structure is* $\mathcal{S}$-*equivalent to one belonging to image* σ.

To prove these results we need an easy sublemma.

3.6.13. Sublemma: *Let* W *be a manifold in* $\mathcal{A}$, *and let* Y *be the submanifold* $(W \times 0) \cup (\partial W \times [0,1]) \subset \partial(W \times [0,1])$. *Let* f *be an* $\mathcal{A}$-*automorphism of* Y *stationary near* $\partial Y = \partial W \times 1$. *Then,* f *extends to an* $\mathcal{A}$-*automorphism* F *of* $W \times [0,1]$ *for which* $F|W \times 1$ *is stationary near* $\partial W \times 1$ *(with respect to any prescribed* $\mathcal{A}$-*collaring).*

Remark: Strictly speaking, the statement of 3.6.13 and its proof below must be slightly adjusted when $\mathcal{A} = \mathcal{D}iff$ to handle corners. For example, we should straighten corners $\partial W \times 0$ and $\partial W \times 1$ via the standard collars of $[0,1]$ and a fixed $\mathcal{A}$-collaring of $\partial W \subset W$, and we should assume f stationary near these corners with respect to the collars used. So much for the statement. We let the reader make the corresponding small adjustment in the proof.

Proof: If $\partial W = \phi$, the result is trivial, so we suppose $\partial W \neq \phi$. The pair $(W \times [0,1], Y)$ is $\mathcal{Q}$-isomorphic to the pair $(W \times [0,1], W \times 0)$. Call the isomorphism h. Now, $g = hf(h^{-1}|W \times 0)$ extends to $g \times id_{[0,1]} = G$, so that $F = h^{-1}Gh$ is an extension of f. The required stationariness of $F| W \times 1$ follows from a corresponding property of G, which is easily checked. Q. E. D.

Remark: F may be chosen to be stationary near $\partial(W \times [0,1])$ with respect to any prescribed $\mathcal{Q}$-collar.

Proof of 3.6.10: We shall make use of Remark 2. of 3.6.8 a), which allows us to use collared $\mathcal{S}$-concordances without loss of generality. We also use 3.3.7, which allows us to assume that f_0 and f_1 are stationary near $M \times \mathbb{R}^{i-1} \subseteq M \times \mathbb{R}^i_+$.

Thus, $\sigma(f_0)$ and $\sigma(f_1)$ are collared structures for $(M_X \times D^i, \partial(M_X \times D^i))$, and there exists a collared $\mathcal{S}$-concordance rel $\partial(M_X \times D^i)$, say H, such that

(i) $H|\partial(M_X \times D^i) \times [0,1] = (\sigma(f_1)|\partial(M_X \times D^i)) \times id_{[0,1]}$

(ii) $H_0 = h\sigma(f_0)$

(iii) $H_1 = \sigma(f_1)$

where $h : M_X \times D^i \to M_X \times D^i$ is an $\mathcal{A}$-automorphism that is stationary near $\partial(M_X \times D^i)$. We shall not make the collars explicit here.

Note that both $\sigma(f_0)$ and $\sigma(f_1)$ have supports disjoint from

$$Z = (M_X \times \{*\}) \cup (\partial M_X \times D^i) \subsetneq \partial(M_X \times D^i).$$

Therefore, $\operatorname{supp}(H|\partial(M_X \times D^i) \times [0,1])$ is disjoint from $Z \times [0,1]$. Since H is collared, this means that $\operatorname{supp} H$ is disjoint from $Z \times [0,1]$. Similarly, one shows that $\operatorname{supp} h$ is disjoint from Z.

Therefore, $h = \sigma(g)$, for some $g \in (\mathcal{A},\mathcal{A})_i(M \text{ rel } X)$, and $H = \sigma(G)$, where G is a $(\mathcal{B},\mathcal{A})$-concordance rel X between gf_0 and f_1. Since every map in $(\mathcal{A},\mathcal{A})_i(M \text{ rel } X)$ is $(\mathcal{B},\mathcal{A})$-concordant rel X^* to the identity, f_0 and f_1 are $(\mathcal{B},\mathcal{A})$-concordant rel X. Q. E. D.

---

*Technically, $(\mathcal{A},\mathcal{A})_i(M \text{ rel } X)$ is not defined because $(\mathcal{A},\mathcal{A})$ is not a pair of geometric categories. That is, in our definition of pair of geometric categories $(\mathcal{B},\mathcal{A})$, we have not allowed $\mathcal{B} = \mathcal{A}$. This is, however, entirely a device for convenient exposition. All of our results, including, for example, the Exactness Theorem, apply to pairs $(\mathcal{A},\mathcal{A})$. In particular, $\pi_i(\mathcal{A},\mathcal{A};M \text{ rel } X) = 0$, for all i. This is what we use here. Alternatively, we could prove the assertion directly, using the methods of the proof of the Exactness Theorem.

Proof of 3.6.11:

Case 1: f_0 or f_1 is a homeomorphism.

Suppose that f_0 is a homeomorphism (and recall that $\mathcal{B} = \mathcal{T}op$ or $\mathcal{H}$). Let F be a $(\mathcal{B},\mathcal{Q})$-concordance rel X between $F_0 = id_{M_X} \times \mathbb{R}^i_+$ and $F_1 = f_1 \circ f_0^{-1}$. We may suppose that F is stationary near $(M_X \times \mathbb{R}^i_+ \times 0) \cup (M_X \times \mathbb{R}^{i-1} \times [0,1]) \cup (M_X \times \mathbb{R}^i_+ \times 1)$ with respect to the standard collars (3.3.7). Define

$$F_X = F | M_X \times \mathbb{R}^i_+ \times [0,1],$$

and let $\sigma(F)$ be the compactification of

$$(id_{M_X} \times \lambda \times id_{[0,1]}) \circ F_X \circ (id_{M_X} \times \lambda^{-1} \times id_{[0,1]}).$$

We shall modify $\sigma(F)$ to make it stationary on $\partial(M_X \times D^i) \times [0,1]$.

Thus, we define the $\mathcal{Q}$-automorphism

$$G = \sigma(F) | (M_X \times D^i \times 0) \cup (\partial(M_X \times D^i) \times [0,1]),$$

and we use 3.6.13 to extend it to $M_X \times D^i \times [0,1]$. Call the extension G, and let

$$g = G | M_X \times D^i \times 1.$$

One easily checks that

$$H = (g \times id_{[0,1]}) \circ G^{-1} \circ \sigma(F)$$

is stationary on $\partial(M_X \times D^i) \times [0,1]$ and that

$$H \circ (\sigma(f_0) \times id_{[0,1]})$$

is an $\mathcal{S}$-concordance rel $\partial(M_X \times D^i)$ between $g \circ \sigma(f_0)$ and $\sigma(f_1 \circ f_0^{-1})\sigma(f_0) = \sigma(f_1)$. This completes Case 1.

Case 2: General case, $i \geq 2$.

When $\mathcal{B} = \mathcal{T}op$, we apply Case 1. When $\mathcal{B} = \mathcal{H}$, we shall modify the situation so that Case 1 can be applied.

Thus, we assume that f_0, f_1 and an $(\mathcal{H}, \mathcal{Q})$-inverse f_2 for f_0 are chosen so that all the sets $\operatorname{supp} f_\alpha$, $\alpha = 0, 1, 2$ are disjoint. (This is where we need $i \geq 2$!) It follows that $f_1 f_2 f_0 = f_2 f_0 f_1$, so that

(1) $$\sigma(f_1)\sigma(f_2)\sigma(f_0) = \sigma(f_2)\sigma(f_0)\sigma(f_1).$$

Now, $\mathrm{id}_{M \times \mathbb{R}^i_+}$ is $(\mathcal{B}, \mathcal{Q})$-concordant rel X to both $f_1 f_2$ and $f_2 f_0$, so that, by Case 1, there are $\mathcal{Q}$-isomorphisms g and h which are $\mathcal{S}$-concordant rel $\partial(M_X \times D^i)$ to $\sigma(f_1)\sigma(f_2)$ and $\sigma(f_2)\sigma(f_0)$, respectively. Let G and H be the corresponding $\mathcal{S}$-concordances, with $G_0 = g$ and $H_0 = h$. Then, $G \circ (\sigma(f_0) \times \mathrm{id}_{[0,1]})$ is an $\mathcal{S}$-concordance between $g\sigma(f_0)$ and

(2) $$\sigma(f_1)\sigma(f_2)\sigma(f_0),$$

and $H \circ (\sigma(f_1) \times \mathrm{id}_{[0,1]})$ is an $\mathcal{S}$-concordance between $h\sigma(f_1)$ and

(3) $$\sigma(f_2)\sigma(f_0)\sigma(f_1).$$

Combining (1)-(3), we see that $h^{-1}g\sigma(f_0)$ is $\mathcal{S}$-concordant to $\sigma(f_1)$. Q.E.D.

Case 3: $\mathcal{B} = \mathcal{H}$, $i = 1$

We use the notation of Case 2. In this case, we cannot suppose that the sets $\operatorname{supp} f_i$, $i = 0, 1, 2$, are disjoint. The maps

$$G \circ (\sigma(f_0) \times id_{[0,1]})$$

and

$$(\sigma(f_1) \times id_{[0,1]}) \circ H$$

provide $\mathcal{S}$-concordances rel boundary between $\sigma(f_1)\sigma(f_2)\sigma(f_0)$ and

$$g\sigma(f_0)$$

and

$$\sigma(f_1)h,$$

respectively. Thus, these two structures are $\mathcal{S}$-concordant rel boundary.

We shall show that, for some $\mathcal{Q}$-isomorphism k, $\sigma(f_1)h$ is $\mathcal{S}$-concordant rel boundary to $k\sigma(f_1)$. Clearly this yields the desired result.

We identify $(D^1, *)$ with $([0,1], 1)$. Note that both $\operatorname{supp} \sigma(f_1)$ and $\operatorname{supp} h$ avoid $M_X \times 1$. We alter $\sigma(f_1)$ and h by an $\mathcal{S}$-concordance rel boundary, if necessary, so that $\sigma(f_1)$ is stationary on $M_X \times [0, \frac{1}{2}]$ and $\operatorname{supp} h \subset M_X \times [0, \frac{1}{2}]$. Define the $\mathcal{Q}$-isomorphism $\ell : M_X \times [0,1] \to M_X \times [0,1]$ by

$$\ell = (\sigma(f_1) | M_X \times 0) \times id_{[0,1]}.$$

Then

$$\sigma(f_1)h = k\sigma(f_1)$$

for $k = \ell h \ell^{-1}$. Q. E. D.

Proof of 3.6.12: Let

$$f : (M^n_X \times D^i, \partial(M^n_X \times D^i)) \to (N^{n+i}, \partial N^{n+i})$$

by a given structure. By Remark 2 of 3.6.8 a), we may assume, without loss of generality, that f is collared. We shall produce an $\mathcal{A}$-isomorphism

$$h : (M^n_X \times D^i, \partial(M^n_X \times D^i)) \to (N^{n+i}, \partial N^{n+i})$$

extending

$$f| (M^n_X \times D^{i-1}_+) \cup (\partial M^n_X \times D^i),$$

where $D^{i-1}_+ = S^{i-1} \cap \mathbb{R}^i_+$. Recall that $* \in \operatorname{int} D^{i-1}_+$ (see 3.6.8 a)).

Suppose that we have found such an h. We then make it stationary near the boundary, using the same collars with respect to which f is stationary. This requires an argument similar to that given for 3.3.7. Now the commutative diagram

$$\begin{array}{ccc}
(M^n_X \times D^i, \partial(M^n_X \times D^i)) & & \\
& \searrow^{f} & \\
\downarrow h^{-1}f & & (N^{n+i}, \partial N^{n+i}). \\
& \nearrow_{h} & \\
(M^n_X \times D^i, \partial(M^n_X \times D^i)) & &
\end{array}$$

shows that the structure f is $\mathcal{S}$-equivalent to $h^{-1}f$. Since $h^{-1}f \in \text{image } \sigma$ (see 3.6.9), we are finished. Therefore, it remains only to find h.

It will be convenient to define D^{i-1}_- not as $S^{i-1} \cap \mathbb{R}^i_-$, but as a slightly shrunk ball contained in the interior, $S^{i-1} \cap \operatorname{int} \mathbb{R}^i_-$. Let

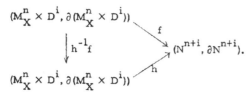

$\bar{M} \subset \operatorname{int} M^n_X$ be obtained by shrinking M^n_X slightly, with respect to some $\mathcal{Q}$-collaring. Finally, let M_+ be obtained from

$$(M^n_X \times D^{i-1}_+) \cup (\partial M^n_X \times D^i) \subsetneq \partial(M^n_X \times D^i)$$

and let M_- be obtained from

$$\bar{M} \times D^{i-1}_- \subsetneq \partial(M^n_X \times D^i)$$

by rounding the corners of these submanifolds-with-corners, if necessary, so that they become $\mathcal{Q}$-submanifolds of $M^n_X \times D^i$. It is not hard to produce an $\mathcal{Q}$-isomorphism

$$(M^n_X \times D^i, M_+, M_-) \xrightarrow{k} (M_+ \times [0,1], M_+ \times 0, M_+ \times 1),$$

such that $k(x) = (x, 0)$ for $x \in M_+$ (cf. diagram).

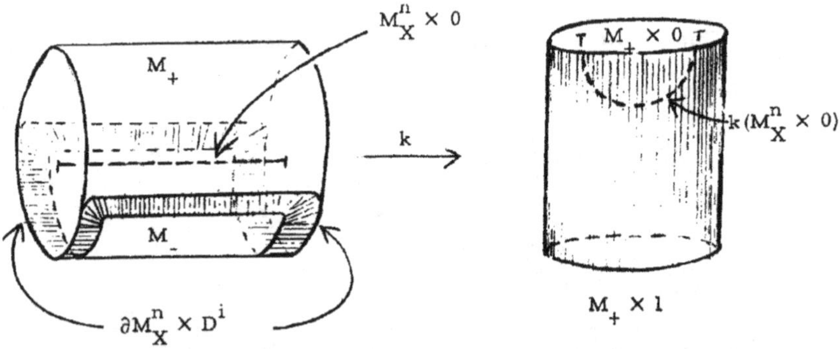

Let $N_{\pm} = f(M_{\pm})$, and choose a homotopy-inverse g for f such that $g|\partial N = f^{-1}|\partial N$. Then, the composite $k \circ g$ shows that (N, N_+, N_-) is a relative h-cobordism between $(N_+, \partial N_+)$ and $(N_-, \partial N_-)$. Moreover, the inclusion

$$N_+ \subset N$$

is precisely the composite

$$N_+ \xrightarrow{g|N_+} M_+ \subseteq M_X^n \times D^i \xrightarrow{f} N.$$

Each of these factor maps is a simple-homotopy equivalence: a) $g|N_+$ is a homeomorphism, so that, by a result of Kirby and Siebenmann [18], it is simple; b) the same argument applies to f when $\mathcal{B} \neq \mathcal{H}$; c) when $\mathcal{B} = \mathcal{H}$, f is simple by assumption; d) the inclusion in the middle is obviously simple.

Thus, the inclusion $N_+ \subset N$ is a simple homotopy equivalence so that, by the relative s-cobordism theorem (for a good statement, see [30]) $k \circ g|N_+$ extends to an $\mathcal{A}$-isomorphism

$$\ell : (N, N_+, N_-) \to (M_+ \times [0,1], M_+ \times 0, M_+ \times 1).$$

Let $h = \ell^{-1} \circ k$. Q. E. D.

Note that this proof makes no special assumption about i (other than $i \geq 1$), but it does require $n+i \geq 6$ in order to apply the s-cobordism theorem.

We summarize the above lemmas.

3.6.14. <u>Corollary: Suppose that $i \geq 1$ and $n+i \geq 6$. Then the function</u>

$$(\mathcal{B},\mathcal{A})_i(M \text{ rel } X) \xrightarrow{\sigma} \{\text{structures on } (M^n_X \times D^i, \partial(M^n_X \times D^i))\},$$

<u>determines a bijection</u>

$$\pi_i(\mathcal{B},\mathcal{A};M \text{ rel } X) \xrightarrow{\sigma_i} \mathcal{S}_i(M \text{ rel } X)\,.$$

We can use the group structure on $\pi_i(\mathcal{B},\mathcal{A};M \text{ rel } X)$ to give $\mathcal{S}_i(M \text{ rel } X)$ a group structure, or we could define the latter directly and prove that σ_i is a homomorphism. Actually, for our purposes, $\mathcal{S}_i$ is just an intermediary between $\pi_i(\mathcal{B},\mathcal{A};M \text{ rel } X)$ and $[\Sigma^i(M/X), B/A]$, and so we are indifferent as to which procedure is used.

3.6.15. <u>The function</u> $\mathcal{S}_i(M \text{ rel } X) \xrightarrow{\eta_i} [\Sigma^i(M/X), B/A]$.

Here, we simply summarize the results about η_i that we need for the Classification Theorem. Definitions of η_i are given in [24], [33], [37] (modulo reversal of arrows in the latter two-recall 3.6.8 a), Remark 1). They insure the commutativity of the following diagram

$$\begin{array}{ccc} \pi_i(\mathcal{B},\mathcal{A};M \text{ rel } X) & \xrightarrow{\kappa_i} & \\ \sigma_i \downarrow & & [\Sigma^i(M/X), B/A] \\ \mathcal{S}_i(\mathcal{B},\mathcal{A};M \text{ rel } X) & \xrightarrow{\eta_i} & \end{array}$$

Note that σ_i^{-1}, $i \geq 1$, is always (for any $n+i$) a well-defined 1-1 correspondence between some subset of $\mathcal{S}_i$ and all of π_i. Thus, the reader may simply view η_i as an extension of $\mathcal{K}_i$ to all structures (a trivial extension when $n+i \geq 6$). We now review the main results on η_i:

a) Lashof and Rothenberg [24] (and Mazur and Hirsch [12]) prove that η_i is a bijection for all i and n, when $(\mathcal{B}, \mathcal{Q}) = (\mathcal{PL}, \mathcal{Diff})$. (Actually, a relative version of the main result stated in [24] is needed, but this can be obtained by the methods of [12] and [24].)

b) Kirby and Siebenmann [18] prove that η_i is a bijection when $n+i \geq 6$ and $(\mathcal{B}, \mathcal{Q}) = (\mathcal{Top}, \mathcal{PL})$. They have obtained similar results in the $(\mathcal{Top}, \mathcal{Diff})$ case, but these are unpublished. We can deduce the $(\mathcal{Top}, \mathcal{Diff})$ case from the $(\mathcal{PL}, \mathcal{Diff})$ and $(\mathcal{Top}, \mathcal{PL})$ cases--at least when $i \geq 2$--by applying the

(i) Exactness Theorem

(ii) naturality of $\mathcal{K}_i$

(iii) Five Lemma.

(We obtain an exact $(\mathcal{Top}, \mathcal{PL}, \mathcal{Diff})$-sequence for (M, X)--for this we need some facts from Appendix C, C. 7--which the $\mathcal{K}_i$ map to the exact sequence obtained by mapping suspensions of M/X into the fibration:

$$PL/O \to TOP/O \to TOP/PL.)$$

c) The exact sequence of part c) of the Classification Theorem, with $\mathcal{S}_i$ in place of π_i, is due to Sullivan [33], [34] and Wall [37]. Thus, it applies to π_i when $n+i \geq 6$.

Corollary 3.6.14 together with the above, complete our proof of the Classification Theorem.

APPENDIX C : The $(\mathcal{PL}, \mathcal{D}iff)$ case

Many of our constructions and definitions involving pairs of geometric categories $(\mathcal{B}, \mathcal{A})$ (e.g., see §3.3) make use of the fact that there exists a forgetful functor $\mathcal{A} \to \mathcal{B}$. In the $(\mathcal{PL}, \mathcal{D}iff)$ case, the absence of such a functor (see [17]) forces us to modify these constructions and definitions.

The tools used in this modification are due to J. Whitehead [39], and to amendments to Whitehead's work by Munkres [29], by Hirsch and Mazur [12], and by Lashof and Rothenberg [24]. We review some basic definitions, assuming the terminology of PL topology as presented, say, in Hudson [16].

Let V be a $\mathcal{D}iff$ manifold and let U be a (possibly empty) $\mathcal{D}iff$ submanifold of V, closed as a subset of V. A piecewise-differentiable (PD) triangulation of (V, U) consists of a pair of simplicial complexes (L, K) (i.e., K is a subcomplex) of L, and an onto homeomorphism $\alpha : L \to V$ such that $\alpha(K) = U$ and $\alpha|\sigma$ is a C^{∞} imbedding for every closed simplex σ in K. When U is empty or has codimension-0, the existence of such an α follows immediately from Whitehead [39]. For the other cases, we can obtain existence by slightly amending Whitehead's argument (as in [29], 10.8), or by applying the far stronger results of Hirsch and Mazur on triangulation of tubular neighborhoods [12]. We touch on questions of uniqueness later.

The cartesian product of two PD triangulations is a PD triangulation, provided one takes appropriate precautions near corners. We shall not concern ourselves with this detail.

Recall that a PL structure for the pair (V, U) means a PL structure for V with respect to which U becomes a PL submanifold of V. A PL structure for (V, U) that contains a PD triangulation of the pair is called compatible with (V, U). Sometimes we shall abbreviate this and simply say that the PL structure is compatible. If $\alpha : (L, K) \to (V, U)$ is a PD triangulation, we may use the notation $(V, U)_\alpha$ to denote the pair of PL manifolds whose underlying spaces coincide with those of V and U and whose PL structures are determined by α. Alternatively, "α", "β", etc., will often denote compatible PL structures for the pair (V, U) without reference to specific representative PD triangulations.

Cartesian products of PL structures have the obvious meaning. The standard PL structures of $\mathbb{R}^i, \mathbb{R}^{i+1}_+, [0,1]$ will be denoted by "1".

When U is empty, we may suppress reference to it and replace the ordered-pair notation by the usual singleton notation.

Let (Z, Y) be a pair of PL manifolds and let (V, U) be as above. A homeomorphism $f : (Z, Y) \to (V, U)$ will be called a PD homeomorphism if $f \circ s : (L, K) \to (V, U)$ is a PD triangulation, for some triangulation $s : (L, K) \to (Z, Y)$ in the PL structure of (Z, Y). Suppose that (V_1, U_1) is another pair of $\mathcal{D}iff$ manifolds such as (V, U), that α is a compatible PL structure for (V, U), and that $f : (V, U)_\alpha \to (V_1, U_1)$ is a

PD homeomorphism. Then, by composing with an appropriate representative triangulation in α, f determines a compatible PL structure for (V_1, U_1) which we denote by $f_*\alpha$. Clearly,

$$(V, U)_\alpha \xrightarrow{f} (V_1, U_1)_{f_*\alpha}$$

is a PL isomorphism. On the other hand, if β is an arbitrary compatible PL structure for (V_1, U_1), then every PL isomorphism

$$(V, U)_\alpha \to (V_1, U_1)_\beta$$

is a PD homeomorphism

$$(V, U)_\alpha \to (V_1, U_1).$$

PD homeomorphisms are stable under right composition by PL homeomorphisms and left composition by diffeomorphisms. They are not, in general, closed under composition or inversion, however. That is to say, they do not form a category. Since this point is the source of most of the technical difficulties, we give an example.

We define homeomorphisms $f, g : \mathbb{R}^2 \to \mathbb{R}^2$ by:

$$f(x, y) = (x, y - \sin x)$$

$$g(x, y) = \begin{cases} (x, y), & y \leq 0 \\ (x+y, y), & y \geq 0 \end{cases}$$

Since f is a diffeomorphism and g is a PL homeomorphism, both are PD homeomorphisms (with respect to the standard PL and C^∞ structures on $\mathbb{R}^2$), as is $f^{-1}g^{-1}$. However, gf is not PD (with respect to these standard structures), because it fails to be differentiable precisely at all points on the graph of the sine function, which is not contained in a 1-dimensional subpolyhedron of (standard PL) $\mathbb{R}^2$.

C.1 The sets $\mathcal{PD}$(M rel X) and ($\mathcal{PL}$, $\mathcal{D}iff$)(M, N rel X).

Our definitions and notation are similar to those of 3.1.6. M is a $\mathcal{D}iff$ manifold, N is a codimension-0 $\mathcal{D}iff$ submanifold of ∂M, closed as a subset of ∂M, and X is a closed subset of M.

Let I be the set of PL structures on M compatible with M, and let $J \subseteq I$ consist of the PL structures on (M, N) compatible with this pair.

For $\alpha \in I$,

$$\mathcal{PD}(M \text{ rel } X)_\alpha$$

consists of all PD homeomorphisms $M_\alpha \to M$ with compact support that avoids X. For $\beta \in J$,

$$(\mathcal{PL}, \mathcal{D}iff)(M, N \text{ rel } X)_\beta$$

consists of all maps

$$f \in \mathcal{PD}(M \text{ rel } X)_\beta$$

for which $f(N) = N$ and $f|U$ is C^∞, for some neighborhood U of N in M (U depending on f). Note that every such f is, of course, a PD homeomorphism $(M, N)_\beta \to (M, N)$, in the sense of the definitions in the above introduction.

Note that, for every $\beta \in J$, there are inclusions

(1)
$$\begin{array}{ccccc} \mathcal{D}\mathit{iff}(M \text{ rel } X) & & \mathcal{PL}(M_\beta \text{ rel } X) & & (\mathcal{PL}, \mathcal{D}\mathit{iff})(M, N \text{ rel } X)_\beta \\ & \subset & \cap & \supset & \\ & & \mathcal{PD}(M \text{ rel } X)_\beta . & & \end{array}$$

We now define the disjoint unions

$$\mathcal{PD}(M \text{ rel } X) = \bigcup_{\alpha \in I} \mathcal{PD}(M \text{ rel } X)_\alpha$$

$$(\mathcal{PL}, \mathcal{D}\mathit{iff})(M, N \text{ rel } X) = \bigcup_{\beta \in J} (\mathcal{PL}, \mathcal{D}\mathit{iff})(M, N \text{ rel } X)_\beta \ .$$

Thus, members of the first set are indexed maps ${}^\alpha f$, where $\alpha \in I$ and $f \in \mathcal{PD}(M \text{ rel } X)_\alpha$. Similarly for the second set. We may abbreviate ${}^\alpha f$ to f when convenient and when no confusion will result.

For each $\beta \in J$, diagram (1) determines inclusions

(2)
$$\begin{array}{ccccc} \mathcal{D}\mathit{iff}(M \text{ rel } X) & & \mathcal{PL}(M_\beta \text{ rel } X) & & (\mathcal{PL}, \mathcal{D}\mathit{iff})(M, N \text{ rel } X) \\ & \subset & \cap & \supset & \\ & & \mathcal{PD}(M \text{ rel } X). & & \end{array}$$

Moreover, there is an obvious inclusion of diagram (1) into diagram (2).

C. 2 Restricted and unrestricted concordance

Choose $\alpha \in I$ and $\beta \in J$. The notions of (restricted) $\mathcal{PD}$-concordance rel X on $\mathcal{PD}(M \text{ rel } X)_\alpha$ and (restricted) $(\mathcal{PL}, \mathcal{D}\mathit{iff})$-concordance rel X on $(\mathcal{PL}, \mathcal{D}\mathit{iff})(M, N \text{ rel } X)_\beta$ are now defined just as their counterparts in 3.1.6. These concordances are maps in

$$\mathcal{PD}(M \times [0,1] \text{ rel } X \times [0,1])_{\alpha \times 1}$$

and
$$(\mathcal{PL}, \mathcal{D}\!\mathit{iff})(M \times [0,1], N \times [0,1] \text{ rel } X \times [0,1])_{\beta \times 1}$$

which are stationary near $\{0,1\}$ (cf. 3.1.6). As before, these concordances determine equivalence relations. Denote the corresponding "factor" <u>sets</u>--recall that there is no composition operation--by

$$\pi_0 \mathcal{PD}; M \text{ rel } X)_\alpha$$

and
$$\pi_0(\mathcal{PL}, \mathcal{D}\!\mathit{iff}; M, N \text{ rel } X)_\beta.$$

The notion of <u>unrestricted concordance</u> rel X is defined for indexed maps in

$$\mathcal{PD}(M \text{ rel } X)$$

and
$$(\mathcal{PL}, \mathcal{D}\!\mathit{iff})(M, N \text{ rel } X).$$

It differs from the above only in that, here, we allow more freedom in the PL structures. Thus, indexed maps

$$^{\alpha_i}f_i \in \mathcal{PD}(M \text{ rel } X), \quad i = 0,1,$$

are <u>unrestricted $\mathcal{PD}$-concordant</u> rel X if and only if there is an indexed map

$$^{\alpha}F \in \mathcal{PD}(M \times [0,1] \text{ rel } X \times [0,1]),$$

with the usual properties of a concordance, where α is a compatible

PL structure for the pair $(M \times [0,1], (M \times 0) \cup (M \times 1))$, such that α coincides with $\alpha_i \times 1$ near (i.e., within some $\varepsilon > 0$) $M \times i$, $i = 0,1$. We express this property of α by saying that α is stationary near $\{0,1\}$ or within ε of $\{0,1\}$.

We define unrestricted $(\mathcal{PL}, \mathcal{D}iff)$-concordance rel X similarly.

These notions define equivalence relations on the corresponding sets, and we denote the corresponding sets of equivalence classes by

$$\pi_0(\mathcal{PD};M \text{ rel } X)$$

and

$$\pi_0(\mathcal{PL}, \mathcal{D}iff;M,N \text{ rel } X).$$

C.3 Some motivation

At this point, we should disclose our reasons for considering PD maps at all and also our reasons for considering indexed maps and the two kinds of concordance defined in C.2.

Choose any $\alpha \in I$, and recall that there are inclusions

$$\begin{array}{ccc} \mathcal{D}iff(M \text{ rel } X) & \to & \mathcal{PD}(M \text{ rel } X)_\alpha \\ & & \uparrow \\ & & \mathcal{PL}(M_\alpha \text{ rel } X). \end{array}$$

In other words, there is a forgetful map from $\mathcal{D}iff$ functions to $\mathcal{PD}$ functions. Lemma 1 in C.4 below will assert that the vertical inclusion defines a bijection

$$\pi_0(\mathcal{PL};M_\alpha \text{ rel } X) \approx \pi_0(\mathcal{PD};M \text{ rel } X)_\alpha$$

Thus, by an "inessential" enlargement of the set of PL maps (to PD maps), we are able to pass from $\mathcal{D}iff$ to $\mathcal{PL}$ as if there were a forgetful functor. This idea has roots in Whitehead's work [39] but was first explicitly used by Hirsch and Mazur [12] and Lashof and Rothenberg [24]. It is expressed in precise functorial form in [17].

This explains our use of PD maps and restricted $\mathcal{PD}$-concordance. In order to define the $(\mathcal{PL}, \mathcal{D}iff)$-sequence for (M, X), we are then led to introduce $(\mathcal{PL}, \mathcal{D}iff)$-maps and restricted $(\mathcal{PL}, \mathcal{D}iff)$-concordance as above. These notions are adequate for most of the subsequent material (e.g., the Exactness Theorem, the First Naturality Theorem, parts b) and c) of the Classification Theorem with $X = \phi$).

There are, however, difficulties with the Second Naturality Theorem and the related portion of part a) of the Classification Theorem. Moreover, the introduction of a group operation for the sets $\pi_i(\mathcal{PL}, \mathcal{D}iff; M \text{ rel } X)_\beta$, $i \geq 1$, involves unnatural choices. Finally, within the framework of restricted concordance classes alone, there seems to be no way of showing that the resulting $(\mathcal{PL}, \mathcal{D}iff)$-sequence for (M, X) is independent of the choice of a particular compatible PL structure for M.

In short, the notion of restricted concordance is too restrictive. All of the above difficulties can be fairly nicely resolved by passing to indexed $(\mathcal{PL}, \mathcal{D}iff)$-maps and unrestricted concordance. For example, a natural (i.e., "composition"-induced) group operation can be introduced into $\pi_0(\mathcal{PL}, \mathcal{Diff}; M, N \text{ rel } X)$, something that, in the restricted case, could only be done--and poorly at that-- for π_i, $i \geq 1$, and $\partial M = \phi$.

We need to know, of course, that the restricted and unrestricted cases are bijectively related. A precise statement to this effect is given in Lemma 2 below.

C.4 The main lemmas

We shall prove these lemmas in Section C.9 of this appendix.

Lemma 1: Choose any $\alpha \in I$. The inclusion

$$\mathcal{PL}(M_\alpha \text{ rel } X) \subset \mathcal{PD}(M \text{ rel } X)_\alpha$$

induces a bijection

$$\pi_0(\mathcal{PL}; M_\alpha \text{ rel } X) \approx \pi_0(\mathcal{PD}; M \text{ rel } X)_\alpha .$$

Via this bijection we can endow $\pi_0(\mathcal{PD}; M \text{ rel } X)_\alpha$ with a group operation.

Lemma 2: Choose any $\alpha \in I$ and $\beta \in J$. The inclusions

$$\mathcal{PD}(M \text{ rel } X)_\alpha \subset \mathcal{PD}(M \text{ rel } X)$$

$$(\mathcal{PL}, \mathcal{Diff})(M, N \text{ rel } X)_\beta \subset (\mathcal{PL}, \mathcal{Diff})(M, N \text{ rel } X)$$

induce bijections

$$\pi_0(\mathcal{PD};M \text{ rel } X)_\alpha \approx \pi_0'(\mathcal{PD};M \text{ rel } X)$$

$$\pi_0(\mathcal{PL},\mathcal{D}iff\,;M,N \text{ rel } X)_\beta \approx \pi_0(\mathcal{PL},\mathcal{D}iff\,;M,N \text{ rel } X).$$

In Section C. 9 we also present a number of technical lemmas to which we occasionally refer below.

C. 5 <u>The groups</u> $\pi_i(\mathcal{PD};M \text{ rel } X)$ <u>and</u> $\pi_{i+1}(\mathcal{PL},\mathcal{D}iff\,;M \text{ rel } X)$

M is now to be a $\mathcal{D}iff$ manifold with empty boundary; X is a closed subset of M; α is some PL structure on M compatible with its $\mathcal{D}iff$ structure. Then, for $i \geq 0$,

$$\mathcal{PD}_i(M \text{ rel } X)_\alpha = \mathcal{PD}(M \times \mathbb{R}^i \text{ rel } X \times \mathbb{R}^i)_{\alpha \times 1}$$

$$(\mathcal{PL},\mathcal{D}iff)_{i+1}(M \text{ rel } X)_\alpha = (\mathcal{PL},\mathcal{D}iff)(M \times \mathbb{R}_+^{i+1}, M \times \mathbb{R}^i \text{ rel } X \times \mathbb{R}_+^{i+1})_{\alpha \times 1}.$$

We also define

$$\mathcal{PD}_i(M \text{ rel } X) = \mathcal{PD}(M \times \mathbb{R}^i \text{ rel } X \times \mathbb{R}^i)$$

$$(\mathcal{PL},\mathcal{D}iff)_{i+1}(M \text{ rel } X) = (\mathcal{PL},\mathcal{D}iff)(M \times \mathbb{R}_+^{i+1}, M \times \mathbb{R}^i \text{ rel } X \times \mathbb{R}_+^{i+1}).$$

Note that the indices for these disjoint unions consist of <u>all</u> compatible PL structures, not just those of the form $\alpha \times 1$ as above.

Passing to concordance classes (restricted, in the first case, unrestricted, in the second), we obtain sets

$$\pi_i(\mathcal{PD};M \text{ rel } X)_\alpha$$

$$\pi_{i+1}(\mathcal{PL},\mathcal{D}iff\,;M \text{ rel } X)_\alpha$$

and

$$\pi_i(\mathcal{PD};M \text{ rel } X)$$

$$\pi_{i+1}(\mathcal{PL}, \mathcal{Diff};M \text{ rel } X).$$

In C.10, we show how to endow these latter sets with a group operation (for all $i \geq 0$). The operation has the property that, for all compatible α on M, the composite bijection

$$\pi_i(\mathcal{PL};M_\alpha \text{ rel } X) \approx \pi_i(\mathcal{PD};M \text{ rel } X)_\alpha \approx \pi_i(\mathcal{PD};M \text{ rel } X)$$

is an isomorphism (C.10, Corollary 2 to Lemma 9). Moreover, there is a notion of juxtaposition which allows us to show that the group operation is abelian when $i \geq 1$. We do this at the end of C.10.

C.6 Exactness

We define sequences of maps

(1) $$\cdots \xrightarrow{\partial} \pi_i(\mathcal{Diff};M) \xrightarrow{j} \pi_i(\mathcal{PD};M)_\alpha \xrightarrow{k} \pi_i(\mathcal{PL}, \mathcal{Diff};M)_\alpha \xrightarrow{\partial} \cdots$$

and

(2) $$\cdots \xrightarrow{\partial} \pi_i(\mathcal{Diff};M) \xrightarrow{j} \pi_i(\mathcal{PD};M) \xrightarrow{k} \pi_i(\mathcal{PL}, \mathcal{Diff};M) \xrightarrow{\partial} \cdots$$

just as in Section 3.3. (X is suppressed for notational ease.) In sequence (2) there is some apparent ambiguity in the definition of j, because it is defined via the composite inclusion,

$$\mathcal{Diff}_i(M) \subseteq \mathcal{PD}_i(M)_\alpha \subseteq \mathcal{PD}_i(M),$$

which depends on α. In C. 9, however, we show that j itself is independent of α. Note that we may replace the middle groups by their PL counterparts via Lemma 1.

The maps in the sequence are easily checked to be homomorphisms (see C. 10).

There is an inclusion-induced homomorphism taking sequence (1) to sequence (2). According to Lemma 1 in C. 4, it is an isomorphism. Thus, we need prove exactness for only one of the sequences. It is more convenient, here, to work with (1), because our proof of the Exactness Theorem in §3. 3, which we try to imitate here, makes implicit use of the product structure on $M \times \mathbb{R}^i$, $M \times \mathbb{R}^i_+$, etc., and the standard structures on $\mathbb{R}^i$, $\mathbb{R}^i_+$, $[0,1]$.

The proof of exactness proceeds just as in section 3. 3. Thus, just as in our proof there, in each of the cases a) - c), we define a certain self-map H of

$$W = (M \times \mathbb{R}^i_+ \times 0) \cup (M \times \mathbb{R}^{i-1} \times [0,1]) \cup (M \times \mathbb{R}^i_+ \times 1),$$

which we must extend to a (restricted) $(\mathcal{PL}, \mathcal{D}iff)$-concordance K on $M \times \mathbb{R}^i_+ \times [0,1]$. To begin, we extend H over a collar of W in $M \times \mathbb{R}^i_+ \times [0,1]$ so that, on the other end of the collar, we still have the same configuration of maps as illustrated in (A) - (C) of 3. 3, but now they are all PL. We then use the $\mathcal{PL}$ case of Sublemma 3. 1. 9 to extend H over all of $M \times \mathbb{R}^i_+ \times [0,1]$.

We illustrate this by sketching case c) of the proof. We have a diagram, as in 3.3, which defines H (we use the notation of the proof of 3.3.2):

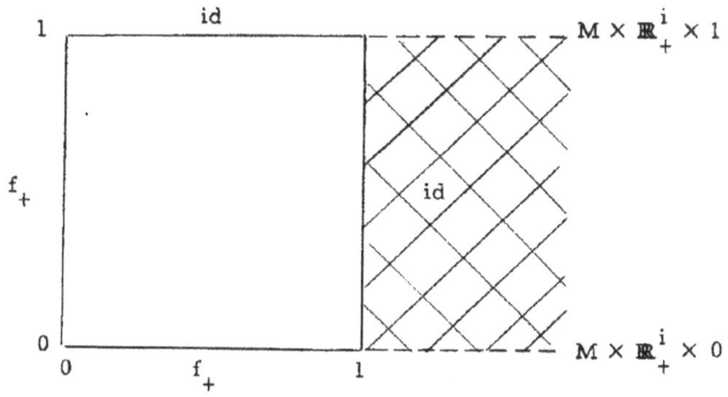

We construct a PD concordance between f_+ and a PL isomorphism g, with the concordance stationary outside some small neighborhood of supp f_+ (see C. 9, Lemma 5'). We use this to fill in regions 1 and 2 below:

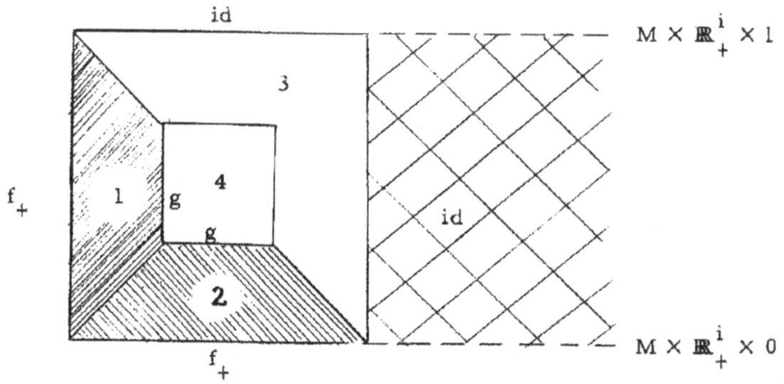

Note that g has compact support. We fill in region 3 with the identity.

Then proceed as in the proof of 3.3.2, using the $\mathcal{PL}$ version of Sub-lemma 3.1.9 to extend over region 4 . Provided the PD isotopy was chosen to be stationary near 0, the result is a restricted ($\mathcal{PL}$, $\mathcal{Diff}$)-concordance. This last condition is easily met.

This completes our discussion of the Exactness Theorem.

C.7 Naturality

We note, first, that we have already stated a special kind of naturality result for the ($\mathcal{PL}$, $\mathcal{Diff}$) case which is not relevant to the cases of pairs of geometric categories treated in the main text: namely, in C.6, above, we asserted that sequence (1) is "included" isomorphically onto sequence (2). We use this fact in our proof, here, of the Second Naturality Theorem.

A relative $\mathcal{Diff}$-map

$$f : (M_1, X_1) \to (M_2, X_2)$$

is defined just as in 3.3.4. We shall define a homomorphism $f^{\#}$ from the ($\mathcal{PL}$, $\mathcal{Diff}$)-sequence for (M_2, X_2) to the ($\mathcal{PL}$, $\mathcal{Diff}$)-sequence for (M_1, X_1). The definition of

$$f^{\#} : \pi_i(\mathcal{Diff}; M_2 \text{ rel } X_2) \to \pi_i(\mathcal{Diff}; M_1 \text{ rel } X_1)$$

is given in 3.3.5.

Now, choose a compatible PL structure α_2 for M, and choose h in $\mathcal{PD}_i(M_2 \text{ rel } X_2)_{\alpha_2}$ or in $(\mathcal{PL}, \mathcal{D}iff)_i(M_2 \text{ rel } X_2)_{\alpha_2}$. We shall define an indexed map $f^{\#}h$ in $\mathcal{PD}_i(M_1 \text{ rel } X_1)$ or in $(\mathcal{PL}, \mathcal{D}iff)_i(M_1 \text{ rel } X_1)$, respectively. In fact, the map itself is defined just as before:

$$f^{\#}h | U \times F = (f^{-1} \times id_F) h (f \times id_F) | U \times F$$
$$f^{\#}h = \text{identity elsewhere,}$$

where U is the open set $f^{-1}(M_2 - X_2)$ described in Definition 3.3.4, and $F = \mathbb{R}^i$ or $\mathbb{R}^i_+$, as the case may be.

We define the index for $f^{\#}h$ (i.e., a compatible PL structure for $M \times F$, F as above), as follows: The diffeomorphism $f^{-1}: M_2 - X_2 \to U \subseteq M_1 - X_1$ transports $\alpha_2 | M_2 - X_2$ to a compatible PL structure α on U. Choose a slightly shrunk open set $U_1 \subset U$ such that supp $f^{\#}h \subset U_1 \times F$, and extend $\alpha | U_1$ arbitrarily to a compatible PL structure on M.* The choices here are, to some extend arbitrary, but we make such choices and fix them for each h. Call the resulting PL structure on M $\alpha(f, h)$. This is the index for h.

As a result, we have maps

$$\mathcal{PD}_i(M_2 \text{ rel } X_2)_{\alpha_2} \to \mathcal{PD}_i(M_1 \text{ rel } X_1)$$

$$(\mathcal{PL}, \mathcal{D}iff)_i(M_2 \text{ rel } X_2)_{\alpha_2} \to (\mathcal{PL}, \mathcal{D}iff)_i(M_1 \text{ rel } X_1).$$

According to C. 9, Lemma 6, the unrestricted concordance class

---

*See C. 9, Lemma 3.

of ${}^{\alpha(f,h)}{}_{f}\#_h$ does not depend on the choice of $U_1 \subset U$ or on the choice of extension of $\alpha | U_1$. That is, the following maps depend only on f:

$$\mathcal{PD}_i(M_2 \text{ rel } X_2)_{\alpha_2} \to \pi_i(\mathcal{PD}; M_1 \text{ rel } X_1)$$

$$(\mathcal{PL}, \mathcal{D}iff)_i(M_2 \text{ rel } X_2)_{\alpha_2} \to \pi_i(\mathcal{PL}, \mathcal{D}iff; M_1 \text{ rel } X_1).$$

If we apply the same construction to unrestricted concordances, we see that the unrestricted concordance class of ${}^{\alpha(f,h)}{}_{f}\#_h$ depends only on f and on the class of h (and not, for example, on α_2). Thus, the above maps induce

$$\pi_i(\mathcal{PD}; M_2 \text{ rel } X_2) \xrightarrow{f^\#} \pi_i(\mathcal{PD}; M_1 \text{ rel } X_1)$$

and
$$\pi_i(\mathcal{PL}, \mathcal{D}iff; M_2 \text{ rel } X_2) \xrightarrow{f^\#} \pi_i(\mathcal{PL}, \mathcal{D}iff; M_1 \text{ rel } X_1).$$

It is not hard to verify that, with these definitions, we obtain the Second Naturality Theorem just as before. The only small matter to check is that $f^\#$ is actually a homomorphism, which is easy (see C.10).

We conclude this section with a brief discussion of change-of-category naturality. Let $(\mathcal{A}, \mathcal{D}iff)$ be either a pair of geometric categories or $(\mathcal{PL}, \mathcal{D}iff)$, and let $(\mathcal{B}, \mathcal{PL})$ be a pair of geometric categories with $\mathcal{B} \geq \mathcal{A}$. Let M be a $\mathcal{D}iff$-manifold and α a compatible PL structure. We shall show how to define "inclusion" homomorphisms

$$\pi_i(\mathcal{A}, \mathcal{D}iff; M \text{ rel } X) \to \pi_i(\mathcal{B}, \mathcal{PL}; M_\alpha \text{ rel } X)$$

which, together with the homomorphisms

$$\pi_i(\mathcal{A};M \text{ rel } X) \to \pi_i(\mathcal{B};M_\alpha \text{ rel } X)$$

defined in 3.3.1 (or, they are the identity homomorphisms when $\mathcal{B} = \mathcal{A}$) and the homomorphisms

$$\pi_i(\mathcal{D}iff;M \text{ rel } X) \to \pi_i(\mathcal{PL};M_\alpha \text{ rel } X)$$

defined in C.5 (using Lemma 1), form a map from the $(\mathcal{A}, \mathcal{D}iff)$-sequence for (M, X) to the $(\mathcal{B}, \mathcal{PL})$-sequence for (M, X).

The idea is to introduce sets

$$(\mathcal{B}, \mathcal{PD})_i(M \text{ rel } X)_\alpha$$

and a relation of (restricted) $(\mathcal{B}, \mathcal{PD})$-concordance rel X, just as before. The concordance classes form

$$\pi_i(\mathcal{B}, \mathcal{PD};M \text{ rel } X)_\alpha .$$

We have inclusions

$$\begin{array}{ccc} (\mathcal{A}, \mathcal{D}iff)_i(M \text{ rel } X) & \to & (\mathcal{B}, \mathcal{PD})_i(M \text{ rel } X)_\alpha \\ & & \uparrow \\ & & (\mathcal{B}, \mathcal{PL})_i(M_\alpha \text{ rel } X). \end{array}$$

The vertical inclusion induces a bijection on the concordance-class level. This is proved in the same way as Lemma 1. Now, in fact, we have the following commutative diagram (in which we delete "M rel X", etc., for simplicity):

$$\begin{array}{ccccc}
\pi_i(\mathcal{D}iff) & \rightarrow & \pi_i(\mathcal{PD}) & \overset{\approx}{\leftarrow} & \pi_i(\mathcal{PL}) \\
\downarrow & & \downarrow & & \downarrow \\
\pi_i(\mathcal{A}) & \rightarrow & \pi_i(\mathcal{B}) & \overset{\text{id}}{\leftrightarrow} & \pi_i(\mathcal{B}) \\
\downarrow & & \downarrow & & \downarrow \\
\pi_i(\mathcal{A}, \mathcal{D}iff) & \rightarrow & \pi_i(\mathcal{B}, \mathcal{PD}) & \overset{\approx}{\leftarrow} & \pi_i(\mathcal{B}, \mathcal{PL}) \\
\downarrow & & \downarrow & & \downarrow \\
\pi_{i-1}(\mathcal{D}iff) & \rightarrow & \pi_{i-1}(\mathcal{PD}) & \overset{\approx}{\leftarrow} & \pi_{i-1}(\mathcal{PL})
\end{array}$$

Thus, to obtain the desired map of exact sequences, we need only reverse the the right-hand arrows.

C.8 The Classification Theorem

The Classification Theorem (Theorem 3.5.5) holds in the $(\mathcal{PL}, \mathcal{D}iff)$ case without change. Some details of the proof must be altered, but, basically, the proof also is the same. We describe, here, the necessary modifications. It will be convenient to work with the group of unrestricted concordance classes $\pi_i(\mathcal{PL}, \mathcal{D}iff; M \text{ rel } X)$.

In Proposition 3.5.3, we assert that the inclusion-induced homomorphism

$$\pi_i(\mathcal{B}, \mathcal{A}; M \text{ rel } V) \rightarrow \pi_i(\mathcal{B}, \mathcal{A}; M \text{ rel } X)$$

is an isomorphism, where V is a regularizing manifold for X. This result also holds for $(\mathcal{B}, \mathcal{A}) = (\mathcal{PL}, \mathcal{D}iff)$, and, in fact, the proof is virtually unchanged. We note here only that the two concordances constructed in that proof

$$K \circ (f \times \text{id}_{[0,1]}) \circ K^{-1} \quad \text{and} \quad G^{-1} \circ H \circ G$$

are unrestricted concordances, in this case, with indices $K_*(\alpha \times 1)$ and $G_*^{-1}\beta$, respectively, where α and β are the indices of f and H respectively.

Just as in 3.6.2, we can reduce the proof of the Classification Theorem to the case in which X is a co-dimension-0, $\mathcal{D}iff$ submanifold of M.

The section on stable differentials remains substantially unaltered. Here, it is important to bear in mind that we are dealing with indexed PD maps ${}^{\alpha}f$, and that the differentials $df : \tau_{M_\alpha} \to \tau_M$ are PD microbundle maps. (The whole theory of PD triangulations and homeomorphisms of manifolds has a microbundle analogue due to Lashof and Rothenberg [24] and a bundle analogue due to Hirsch and Mazur [12]. See also [17].)

The map $\mathcal{K}_i$ (3.6.4, 3.6.5) is defined as before. Details for this construction (in a slightly less general setting) can be found in [24].

Naturality (3.6.6) is proved as before.

For our definition of $\mathcal{S}_i(\mathcal{PL}, \mathcal{D}iff\,; M \text{ rel } X)$ (3.6.7), we repeat the restricted-unrestricted dichotomy used above. Thus, for example, using the notation of 3.6.7 a) and letting α be a compatible PL structure for $(V, \partial V)$, we define a $(\mathcal{PL}, \mathcal{D}iff)$-structure for $(V, \partial V)_\alpha$ to be a map of pairs,

$$(V, \partial V) \xrightarrow{f} (N, \partial N)$$

such that

(i) N is a compact $\mathcal{D}iff$ manifold;

(ii) $f|U$ is a $\mathcal{D}iff$ imbedding for some neighborhood U of ∂V;

(iii) $f : (V, \partial V)_\alpha \to (N, \partial N)$ is a PD homeomorphism.

We define restricted $\mathcal{S}$-equivalence in the obvious way, and obtain $\mathcal{S}(\mathcal{PL}, \mathcal{D}iff; V, \partial V)_\alpha$ as before. Among indexed structures, we can define unrestricted equivalence, obtaining $\mathcal{S}(\mathcal{PL}, \mathcal{D}iff; V, \partial V)$. The "inclusion"

$$\mathcal{S}(\mathcal{PL}, \mathcal{D}iff; V, \partial V)_\alpha \to \mathcal{S}(\mathcal{PL}, \mathcal{D}iff; V, \partial V)$$

is a bijection. This is proved in the same way as Lemma 1 of C.4.

We now prove Lemmas 3.6.9 - 3.6.11 in much the same way as before, using the unrestricted objects. This gives us sufficient freedom to virtually mimick the given proofs. We let the reader supply details.

The function η_i (3.6.14) is defined in the literature (e.g., [24]) for the restricted case. This is because the viewpoint there is slightly different: one begins, not with a $\mathcal{D}iff$ manifold but with a fixed $\mathcal{PL}$ manifold with preferred $\mathcal{D}iff$ structure. Of course, this causes no difficulties for us in view of the bijective connection between the restricted and unrestricted cases. It does suggest, however, another reason for considering both.

This completes our discussion of the Classification Theorem.

C. 9 Some technical results

We shall appeal to ideas and results stated in Munkres [29] and in Hirsch [11], both of which, in turn, depend heavily on Whitehead [39].

Lemma 3: *Let* V *be a* C^∞ *manifold containing open sets* U_0, U_1 *with* $\bar{U}_0 \subset U_1$. *Suppose that* β *is a PL structure on* U_1 *compatible with its* C^∞ *structure. Then,* $\beta | U_0$ *extends to a PL structure on* V *compatible with its* C^∞ *structure.*

This is a "relative" version of Whitehead's theorem on existence of triangulations [39]. It can be proved by using Theorem 10.4 of [29], modified as indicated in Exercise (a), p. 101. Alternatively, it can be proved using the following:

Theorem (*Whitehead* [39]): *Let* W *be a* C^∞ *manifold. Every compatible PL structure on* ∂W *extends to one on* W.

Now choose a PL submanifold $W' \subset (U_1)_\beta$, closed as a subset of V, with $\bar{U}_0 \subset \text{int } W'$. By a PD isotopy that is stationary outside a small bicollar of $\partial W'$, deform $\partial W'$ so that it becomes a smooth submanifold of V (see Hirsch [13]). Let h_1 be the end of the isotopy. Then, $h_1(W\ ) = W$ is a PL submanifold of $(U_1)_{(h_1)_*(\beta)}$, as well as a smooth submanifold of U_1. Note that if the bicollar of $\partial W'$ is small enough then $(h_1)_*(\beta) | U_0 = \beta | U_0$. Now use Whitehead's theorem to extend

$(h_1)_*(\beta)$ over V-int W. This proves Lemma 3.

Let K be a PL manifold, and let Map(K, V) be the set of all onto maps $(K, \partial K) \to (V, \partial V)$ that are PD (i.e., for each such map h, there is a triangulation $s : L \to K$ in the PL structure of K, such that $h \circ s/\sigma$ is C^∞, for each σ in L). Munkres [29] shows how to endow Map(K, V) with a "C^1-topology". Hirsch's formulation in [11] is slightly different, but the resulting topology is the same. We need the following fact about Map (K, V), which follows from results in [29]:

Lemma 4: *In the C^1-topology, the PD homeomorphisms $K \to M$ form an open subset of Map (K, M).*

We now state an important fact, due to Whitehead [39] and Hirsch [11]. It is a special case of Theorem 13.4 of Hirsch [11]:

Lemma 5: *Let $f_i : K_i \to V$, $i = 0, 1$, be PD homeomorphisms, let $A \subset K_0$ be a subpolyhedron with $\overline{K_0 - A}$ compact, and let W be a C^1-neighborhood of f_0 in Map (K_0, V). Suppose, furthermore, that $f_1^{-1} f_0 : A \to K_1$ is PL. Then, there exists a PD homeomorphism h in W that extends $f_0 | A$ such that $f_1^{-1} h$ is PL.*

The requirement that $\overline{K_0 - A}$ be compact causes some inconvenience in our subsequent proofs. As far as we know, no relative result of this type can avoid the condition or some equivalent: it seems to be firmly entrenched in the known proofs (cf. [29], p. 90, Unsolved problem).

Addendum to Lemma 5: Under the hypotheses of Lemma 5, h may be chosen to satisfy the following additional condition: h is (restricted) PD concordant "rel A" to f_0.

By this we mean that there is a PD homeomorphism

$$H : K \times [0,1] \to V \times [0,1]$$

satisfying $H(x,0) = (f_0(x),0), H(x,1) = (h(x),1)$, and $H(y,t) = (f(y),t)$, for $y \in A$.

Proof: Assume first that $\partial V = \phi$. Smoothly imbed V in some Euclidean space, regarding f_0 and h, then, as maps into this Euclidean space. Let T be a smooth tubular neighborhood of V with smooth retraction $r : T \to V$. Define H by

$$H(x,t) = (r(f_0(x) + t(h(x)-f_0(x))),t).$$

One easily checks that this is a well-defined PD isotopy through homeomorphisms, provided h is chosen C^1-close enough to f. By Lemma 5, such a choice is possible. Adjust this isotopy to make it stationary near $\{0,1\}$.

We use essentially the same argument when $\partial V \neq \phi$, except that now we smoothly imbed V so that ∂V lies in a hyperplane and V lies on one side. T is chosen to be a relative tubular neighborhood contained in the half-space containing V. The argument now proceeds as before.

Q. E. D.

We restate Lemma 5, as amended above, in a slightly more convenient form:

Lemma 5': Let f_0, K_0, A and V be as in Lemma 5, and let β be a compatible PL structure for V^* such that $f_0|A : A \to V_\beta$ is PL. Then, there is a PD concordance, stationary on A, between f_0 and a PL homeomorphism $h : K_0 \to V$.

Proof of Lemma 1: We recall that M is a C^∞ manifold and α is a compatible PL structure.

Let $f : M_\alpha \to M$ be a PD homeomorphism with support that is compact and avoids X. Let A be any subpolyhedron of M_α satisfying:

a) M-A has compact closure.

b) $\text{supp} f \cap A = \phi$.

c) $X \subset \text{int}\, A$.

Clearly such A exist. Note that f/A is the inclusion $A \subset M$, which is certainly a PL map $A \to M_\alpha$. Lemma 5' implies that f is $\mathcal{PD}$-concordant rel X to a map $h \in \mathcal{PL}(M_\alpha \text{ rel } X)$. Therefore, the inclusion-induced map

---

*i.e., β is compatible with the C^∞ structure of V.

$$\pi_0(\mathcal{PL};M_\alpha \text{ rel } X) \to \pi_0(\mathcal{PD};M \text{ rel } X)_\alpha$$

is onto.

The proof that the map is one-one is similar. Thus, let $F : (M \times [0,1])_{\alpha \times 1} \to M \times [0,1]$ be a $\mathcal{PD}$-concordance rel X, with F_0 and F_1 PL as maps $M_\alpha \to M_\alpha$.* Let A' be any subpolyhedron of M_α satisfying:

a) M-A' has compact closure

b) supp $F \cap (A' \times [0,1]) = \phi$

c) $X \subset \text{int } A'$.

Again it is easy to see that such A' exist. Let

$$A = (M \times [0,\varepsilon]) \cup A' \times [0,1] \cup (M \times [1-\varepsilon,1]).$$

Then, A is a subpolyhedron of $(M \times [0,1])_{\alpha \times 1}$ with relatively compact complement, and $F|A$ is PL. Let $H : (M \times [0,1])_\alpha \to (M \times [0,1])_\alpha$ be the PL homeomorphism produced by Lemma 5'. Since it extends $F|A$, H is a $\mathcal{PL}$-concordance rel X between F_0 and F_1. Q. E. D.

We now use the notation and terminology introduced in C.1 and in C.2. The following lemma will allow us to sidestep the compactness condition of Lemma 5:

Lemma 6: Suppose that $f \in (\mathcal{PL}, \mathcal{D}iff)(M, N \text{ rel } X)_\beta$ is a C^∞ imbedding on an open neighborhood U of N in M. Let $\beta' \in J$ be any PL

---

*As usual, F is stationary within ε of $\{0,1\}$.

structure coinciding with β on a neighborhood of M-U.

Then, ${}^{\beta}f$ and ${}^{\beta'}f$ are unrestricted ($\mathcal{PL}$, $\mathcal{D}iff$)-concordant rel X.

Proof: The concordance is

$$(M \times [0,1])_{\gamma} \xrightarrow{f \times id} M \times [0,1],$$

where γ is a compatible, PL structure on $(M \times [0,1], N \times [0,1])$ satisfying

a) $\gamma | M \times 0 = \beta$, $\quad \gamma | M \times 1 = \beta'$

b) $\gamma = \beta \times 1$ on a neighborhood of $(M-U) \times [0,1]$. Note that because $f \times id$ is C^{∞} on $U \times [0,1]$, it is, indeed, a PD homeomorphism for any γ satisfying a) and b). It remains to construct γ. When $N = \phi$, properties a) and b), together with Lemma 3 give a suitable γ immediately. When $N \neq \phi$, some care must be exercised to assure that γ be compatible with the pair $(M \times [0,1], N \times [0,1])$. We give some details.

Let W_0 and W_1 be neighborhoods of M-U and N, respectively, such that $W_0 \cap W_1 = \phi$. Define γ^1 by:

a') $\gamma^1 | M \times [0,\frac{1}{3}) = \beta \times 1 | M \times [0,\frac{1}{3})$

$\gamma^1 | M \times (\frac{2}{3},1] = \beta' \times 1 | M \times (\frac{2}{3},1]$

b') $\gamma^1 | W_0 \times [0,1] = \beta \times 1 | W_0 \times [0,1]$.

Note that $\gamma^1 | N \times [0,\frac{1}{3}) \cup N \times (\frac{2}{3},1]$ is indeed a compatible PL structure for this manifold, because β and β' are compatible with the pair (M, N).

Using Lemma 3, we extend $\gamma^1 | N \times [0,\frac{1}{4}) \cup N \times (\frac{3}{4},1]$ arbitrarily to $N \times [0,1]$. Denote by γ^2 the resulting PL structure on $M \times [0,\frac{1}{4}) \cup (W_0 \cup N) \times [0,1] \cup M \times (\frac{3}{4},1]$.

Now consider the smooth manifold $W_1 \times [\frac{1}{8},\frac{7}{8}]$ with boundary $(W_1 \times \frac{1}{8}) \cup \partial W_1 \times [\frac{1}{8},\frac{7}{8}] \cup (W_1 \times \frac{7}{8})$. This boundary contains the smooth submanifold with corners $(W_1 \times \frac{1}{8}) \cup N \times [\frac{1}{8},\frac{7}{8}] \cup W_1 \times \frac{7}{8}$, which γ^2 endows with a compatible PL structure. Use Whitehead's theorem to extend this structure to the entire boundary, and then use Whitehead's theorem again to extend the result to $W_1 \times [\frac{1}{8},\frac{7}{8}]$.

We now have a PL structure γ^3 defined on $M \times [0,\frac{1}{8}) \cup (W_0 \cup W_1) \times [0,1] \cup M \times (\frac{7}{8},1]$ and satisfying a) and b), above. Shrink its domain slightly, and extend the result arbitrarily to $M \times [0,1]$. This is γ. Q. E. D.

In C. 2, we assert that the inclusion-induced composite map

$$\pi_i(\mathcal{D}iff;M \text{ rel } X) \to \pi_i(\mathcal{PD};M \text{ rel } X)_\alpha \to \pi_i(\mathcal{PD};M \text{ rel } X)$$

is independent of the choice of compatible PL structure $\alpha \in I$. The proof of this, now, is immediate. For, let α_1, α_2 be any two such structures, and choose any $f \in \mathcal{D}iff_i(M \text{ rel } X)$. Then, ${}^{\alpha_1 \times 1}f$ and ${}^{\alpha_2 \times 1}f$ are the two indexed maps representing the images of the class of f under the "two" homomorphisms. Since f is C^∞ on all of $M \times \mathbb{R}^i$, however, Lemma 6 implies that the two indexed maps are unrestricted $\mathcal{PD}$-concordant rel X, as desired.

Lemma 7: Choose any $\beta_1, \beta_2 \in J$ and any $f \in (\mathcal{PL}, \mathcal{Diff})(M, N \text{ rel } X)_{\beta_1}$. Then, there is a $g \in (\mathcal{PL}, \mathcal{Diff})(M, N \text{ rel } X)_{\beta_2}$ that is unrestricted $(\mathcal{PL}, \mathcal{Diff})$-concordant to f.

Proof: Let W_1 and W_2 be open subsets of M with the following properties (see the diagram below):

a) $\overline{W}_1 \cap \overline{W}_2 = \phi$.

b) f is C^∞ on a neighborhood of $M - W_1$.

c) $N \cup X \subset W_2$.

d) $M - W_2$ is compact.

Let β be any PL structure for (M,N), compatible with its C^∞ structure and extending $\beta_1 | W_1 \cup \beta_2 | W_2$.* According to Lemma 6, ${}^{\beta_1}f$ is $(\mathcal{PL}, \mathcal{Diff})$-concordant rel X to ${}^{\beta}f$.

Now, let $b : K \to M$ be a PD triangulation representing β, and let $A \subset K$ be a subpolyhedron such that $M - b(A)$ is a relatively compact neighborhood of $M - W_2$ and $N \cup X \subset b(A)$. According to Lemma 5', there is a PD concordance $H : K \times [0,1] \to M \times [0,1]$ such that

e) $H_0 = b$

f) $H_1 : K \to M_{\beta_2}$ is PL

g) H is stationary on A.

Let $G = (f \times id_{[0,1]}) \circ (b \times id_{[0,1]}) \circ H^{-1}$, and let γ be the PL structure

*One need only choose the extension β to be compatible with M, because $N \subset W_2$ and β_2 is compatible with (W_2, N), by hypothesis.

with which H endows $M \times [0,1]$. Then, $G : (M \times [0,1])_\gamma \to M \times [0,1]$ is an unrestricted $(\mathcal{PL}, \mathcal{D}iff)$-concordance rel X between ${}^{\beta}f$ and ${}^{\beta_2}g$, where $g = f \circ b \circ H_1^{-1}$. Q. E. D.

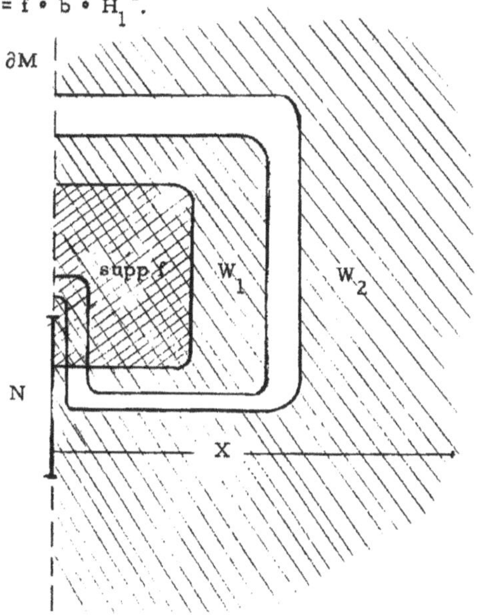

By a slight refinement of the proof of Lemma 7, we can obtain the following:

Lemma 8: *Let* $f_i \in (\mathcal{PL}, \mathcal{D}iff)(M, N \text{ rel } X)_{\beta_i}$, $i = 0,1$ *be unrestricted* $(\mathcal{PL}, \mathcal{D}iff)$*-concordant* rel X, *and suppose that* γ *is any compatible* PL *structure for* $(M \times [0,1], N \times [0,1])$ *extending* $(\beta_0 | M \times 0) \cup (\beta_1 | M \times 1)$. *Then, there exists an unrestricted* $(\mathcal{PL}, \mathcal{D}iff)$*-concordance* $G \in (\mathcal{PL}, \mathcal{D}iff)(M \times [0,1], N \times [0,1] \text{ rel } X \times [0,1])_\gamma$ *with* $G_i = f_i$, $i = 0,1$.

Proof: Let

$$F \in (\mathcal{PL}, \mathcal{Diff})(M \times [0,1],\ N \times [0,1] \text{ rel } X \times [0,1])_\delta$$

be the postulated unrestricted concordance between f_0 and f_1. We must find an unrestricted concordance G between these maps such that

$$G \in (\mathcal{PL}, \mathcal{Diff})(M \times [0,1],\ N \times [0,1] \text{ rel } X \times [0,1])_\gamma .$$

This G is constructed just as is the g of Lemma 7. The only additional condition that G must satisfy here is that it extend $F|(M \times [0,\varepsilon]) \cup (M \times [1-\varepsilon, 1])$, for some $\varepsilon > 0$. Let $B : K \times [0,1] \to M \times [0,1]$ be the analogue of the b in the proof of Lemma 7, with B stationary within ε of $\{0,1\}$, and let A' be the analogue of the polyhedron A in that proof. Then

$$A = B^{-1}(M \times [0,\varepsilon]) \cup A' \cup B^{-1}(M \times [1-\varepsilon, 1])$$

is a subpolyhedron of $K \times [0,1]$, and

$$B|A : A \to (M \times [0,1])_\gamma$$

is PL. Now obtain H as before, H stationary on A. Then the desired map G is $F \circ B \circ H_1^{-1}$. Q. E. D.

Proof of Lemma 2: We mention only the $(\mathcal{PL}, \mathcal{Diff})$ case: the other follows by specializing $N = \phi$. Lemma 7 implies that the inclusion-induced map

$$\pi_0(\mathcal{PL}, \mathcal{Diff}; M, N \text{ rel } X)_\beta \to \pi_0(\mathcal{PL}, \mathcal{Diff}; M, N \text{ rel } X)$$

is onto. Lemma 8 implies that it is 1-1. Q. E. D.

C.10. The group operations

Again we shall discuss only the more general $(\mathcal{PL}, \mathcal{D}iff)$-case.

We begin by defining a composition operation for certain pairs of indexed maps in

$$(\mathcal{PL}, \mathcal{D}iff)(M, N \text{ rel } X).$$

We consider only pairs $({}^{\alpha}f, {}^{\beta}g)$ of such maps for which

(1) $$\alpha = g_*\beta.$$

The composite, then, is given by

(2) $$({}^{\alpha}f) \circ ({}^{\beta}g) = {}^{\beta}(f \circ g).$$

Assuming for the moment that this partial composition well-defines an operation all of

$$\pi_0(\mathcal{PL}, \mathcal{D}iff; M, N \text{ rel } X),$$

we note that the equations

(3) $$({}^{g_*\beta}\mathrm{id}_M) \circ ({}^{\beta}g) = ({}^{\beta}g) \circ ({}^{\beta}\mathrm{id}_M) = {}^{\beta}g$$

(4) $$({}^{g_*\beta}g^{-1}) \circ ({}^{\beta}g) = {}^{\beta}(\mathrm{id}_M); \ ({}^{\beta}g) \circ ({}^{g_*\beta}g^{-1}) = {}^{g_*\beta}(\mathrm{id}_M).$$

(5) $$(({}^{\alpha}f) \circ ({}^{\beta}g)) \circ ({}^{\gamma}h) = {}^{\gamma}(f \circ g \circ h) = ({}^{\alpha}f) \circ (({}^{\beta}g) \circ ({}^{\gamma}h)),$$

(where $\alpha = g_*\beta$ and $\beta = h_*\gamma$),

imply that the operation is a group operation.

Lemma 9: Let ${}^{\alpha}f, {}^{\alpha'}f', {}^{\beta}g, {}^{\beta'}g'$ be as above such that ${}^{\alpha}f$ (resp., ${}^{\beta}g$) is unrestricted concordant to ${}^{\alpha'}f'$ (resp., ${}^{\beta'}g'$) and $\alpha = g_*\beta$ (resp., $\alpha' = g'_*\beta'$). Then, ${}^{\beta}(f \circ g)$ is unrestricted concordant to ${}^{\beta'}(f' \circ g')$.

Proof: Let ${}^{\delta}G$ be an unrestricted concordance between ${}^{\beta}g$ and ${}^{\beta'}g'$, and choose an unrestricted concordance ${}^{\gamma}F$ between ${}^{\alpha}f$ and ${}^{\alpha'}f'$, with $\gamma = G_*\delta$. According to Lemma 8, such a choice is always possible. Then ${}^{\delta}(F \circ G)$ is an unrestricted concordance between ${}^{\beta}(f \circ g)$ and ${}^{\beta'}(f' \circ g')$.

Q. E. D.

Corollary 1 to Lemma 9: The partial composition given by equation (2), subject to condition (1), well-defines a group operation on

$$\pi_0(\mathcal{PL}, \mathcal{Diff}; M, N \text{ rel } X).$$

Proof: For any classes $[{}^{\gamma}f]$ and $[{}^{\delta}g]$, let

$$[{}^{\gamma}f] \circ [{}^{\delta}g] \equiv [{}^{\alpha}f' \circ {}^{\beta}g'],$$

where $\alpha = g'_*(\beta)$, ${}^{\alpha}f'$ is unrestricted concordant to ${}^{\gamma}f$ and ${}^{\beta}g'$ is unrestricted concordant to ${}^{\delta}g$. That such ${}^{\alpha}f'$ and ${}^{\beta}g'$ exist follows from Lemma 7. That the resulting operation is well-defined follows from Lemma 9.

Q. E. D.

Corollary 2 to Lemma 9: For every $\alpha \in I$, the composite of inclusion-induced bijections

$$\pi_0(\mathcal{PL}; M_\alpha \text{ rel } X) \approx \pi_0(\mathcal{PD}; M \text{ rel } X)_\alpha \approx \pi_0(\mathcal{PD}; M \text{ rel } X)$$

is a group isomorphism.

Proof: The partial composition operation on $\mathcal{PD}$(M rel X) coincides with ordinary composition on the subset

$$\mathcal{PL}(M_\alpha \text{ rel } X) \subset \mathcal{PD}(M \text{ rel } X)_\alpha \subset \mathcal{PD}(M \text{ rel } X). \qquad \text{Q.E.D.}$$

This shows that the operation on $\pi_0(\mathcal{PD}; M \text{ rel } X)$ induced from that of $\pi_0(\mathcal{PL}; M_\alpha \text{ rel } X)$ via the above bijections is independent of α. For every $\beta \in J$, we now endow

$$\pi_0(\mathcal{PL}, \mathcal{D}iff\,; M, N \text{ rel } X)_\beta$$

with the group operation of

$$\pi_0(\mathcal{PL}, \mathcal{D}iff\,; M, N \text{ rel } X)$$

(pulling it back via the bijection of Lemma 2). It is easy to see that the restricted and unrestricted $(\mathcal{PL}, \mathcal{D}iff)$ sequences* for M rel X are sequences of homomorphisms and that the forgetful map linking them consists of isomorphisms.

We conclude with some remarks about juxtaposition of indexed maps in

$$\mathcal{PD}_i(M \text{ rel } X) \quad \text{or in}$$

$$(\mathcal{PL}, \mathcal{D}iff)_{i+1}(M \text{ rel } X),$$

$i \geq 1$. We restrict attention to the first set; the other is handled

---

*See C.6.

similarly. Note that the indices of maps in $\mathcal{PD}_i(M \text{ rel } X)$ range over all compatible PL structures, not just those of the form $\alpha_0 \times 1$, where α_0 is a structure on M and 1 the standard structure on $\mathbb{R}^i$.

Given any ${}^{\alpha}f$ and any index γ, we claim that ${}^{\alpha}f$ is (unrestricted) $\mathcal{PD}$-concordant rel X to an i-positive indexed map ${}^{\gamma}f_+$ (cf. 3.2.4). To see this, first make ${}^{\alpha}f$ concordant to an indexed map ${}^{\beta}g$, where β is of the form $\beta_0 \times 1$ described above (Lemma 7). Then, apply the argument of Lemma 3.2.5 to make ${}^{\beta}g$ (restricted) concordant to the i-positive map ${}^{\beta}g_+$.* For the last step, we apply Lemma 7 again, only now we note the additional fact that the resulting map with index γ can be chosen to approximate g_+ arbitrarily well. In particular, it can be chosen to be i-positive. This is ${}^{\gamma}f_+$. Of course, a similar argument yields an i-negative map ${}^{\gamma}f_-$.

Now, given any indexed maps ${}^{\alpha}f$ and ${}^{\beta}g$, we define a juxtaposition of ${}^{\alpha}f$ with ${}^{\beta}g$ to be any composition $({}^{\gamma}f_+) \circ ({}^{\delta}g_-)$ such that

a) $(g_-)_*(\delta) = \gamma$;

b) ${}^{\gamma}f_+$ is an i-positive map unrestricted concordant to ${}^{\alpha}f$;

c) ${}^{\delta}g_-$ is an i-negative map unrestricted concordant to ${}^{\beta}g$.

The argument above shows that such ${}^{\gamma}f_+$ and ${}^{\delta}g_-$ can always be found.

Note that because $(g_-)_*(\delta) = \gamma$ and ${}^{\gamma}f_+$ is i-positive, δ is an

---

*This argument requires β to have the form $\beta_0 \times 1$, since we conjugate with a map of the form $id_M \times A$, A affine.

admissible index for f_+. Similarly, $\gamma' = (f_+)_*(\delta)$ is an admissible index for g_-.

Suppose now that $\alpha = g_*(\beta)$, so that the composition $({}^{\alpha}f) \circ ({}^{\beta}g)$ is defined. We have, then, the following chain of unrestricted concordances:

$$
\begin{aligned}
({}^{\alpha}f) \circ ({}^{\beta}g) \sim ({}^{\gamma}f_+) \circ ({}^{\delta}g_-) &= {}^{\delta}(f_+g_-) \\
&= {}^{\delta}(g_-f_+) = ({}^{\gamma'}g) \circ ({}^{\delta}f_+).
\end{aligned}
$$

Therefore, when juxtaposition is defined, that is when $i \geq 1$, *the group operations induced in* $\pi_i(\mathcal{PD}; M \text{ rel } X)$ *and in* $\pi_{i+1}(\mathcal{PL}, \mathcal{D}iff; M \text{ rel } X)$ *are abelian.*

APPENDIX D: Computing $\pi_i(\mathcal{H};M \text{ rel } X)$

In this appendix, we reduce the problem of computing

$$\pi_i(\mathcal{H};M \text{ rel } X), \quad i \geq 1,$$

to that of computing the homotopy groups of a certain function space. Under certain conditions--for example, those described in Proposition 3.4.2--these homotopy groups are isomorphic to

$$[\Sigma^i(M/X), M-X],$$

the group of based-homotopy classes of based-maps from the i-fold reduced suspension $\Sigma^i(M/X)$ to M-X. Here, M/X is the usual reduced mapping cone of the inclusion $X \subseteq M$, which equals $M \cup$ basepoint when X is empty.

From now on, *we assume that* M *is closed and that* X *is regular in* M (see 3.5.1). V will denote a regularizing manifold for X, and M_V will denote M-int V.

For spaces A and B, let Map(A, B) be the space of continuous maps $A \to B$ endowed with the compact-open topology.

D.1 *Lemma: When* A *is a compact metric space and* B *is a separable metric* ANR, *then* Map(A, B) *is a separable, metric,* ANR.

This is a result of Kuratowski [23], p. 284.

Let A' be a closed subset of A. Then, there is a map, defined by restriction to A',

$$p : \mathrm{Map}(A, B) \to \mathrm{Map}(A', B).$$

D. 2 Lemma: p is a Hurewicz fibration. (See [31], p. 97.)

D. 3 Corollary: The fibres of p are separable metric ANR's.

Proof: It suffices to prove that the fibres are ANR's.

If $\mathrm{Map}(A', B)$ is contractible, then $\mathrm{Map}(A, B)$ has the fibre-homotopy type of $p^{-1}(x) \times \mathrm{Map}(A', B)$, and so the inclusion $p^{-1}(x) \subseteq \mathrm{Map}(A, B)$ is a homotopy equivalence. Since $p^{-1}(x)$ is a closed subset of $\mathrm{Map}(A, B)$, a result of Fox [8] implies that it is a strong deformation retract. Thus, because $\mathrm{Map}(A, B)$ is an ANR, so is $p^{-1}(x)$.

In general, $\mathrm{Map}(A', B)$, being an ANR, is locally contractible. Therefore, since open subsets of ANR's are ANR's, we may apply the above argument to the fibration $p : p^{-1}(U) \to U$, where U is an open, contractible neighborhood of $x \in \mathrm{Map}(A', B)$. Q. E. D.

Let $g : \partial V \to M_V = M\text{-int } V$ be any continuous map, and define

$$\mathcal{M}(M_V, g)$$

to be the fibre $p^{-1}(g)$ in the fibration

$$\mathrm{Map}(M_V, M_V) \xrightarrow{p} \mathrm{Map}(\partial V, M_V).$$

D. 4 Corollary: $\mathcal{M}(M_V, g)$ is a separable, metric ANR.

Let $\iota : \partial V \to M_V$ be the inclusion map. Then, every class in

$$\pi_i(\mathcal{M}(M_V, \iota), \mathrm{id}_{M_V})$$

can be represented by a map

$$f : M_V \times \mathbb{R}^i \to M_V \times \mathbb{R}^i$$

preserving the second factor and satisfying

a) $f | \partial V \times \mathbb{R}^i$ is the inclusion $\partial V \times \mathbb{R}^i \subseteq M_V \times \mathbb{R}^i$.

b) $f(x, y) = (x, y)$, for $|y| \geq 1$.

Extend such an f to

$$\bar{f} : M \times \mathbb{R}^i \to M \times \mathbb{R}^i$$

by defining it to be the identity on $V \times \mathbb{R}^i$.

D.5 <u>Lemma:</u> a) $\bar{f} \in \mathcal{H}_i(M \text{ rel } X)$

b) <u>The correspondence</u> $f \to \bar{f}$ <u>defines a homomorphism</u>

$$\pi_i(\mathcal{M}(M_V, \iota), \mathrm{id}_{M_V}) \xrightarrow{\Phi_V} \pi_i(\mathcal{H}; M \text{ rel } X).$$

<u>Proof:</u> a) Let g represent the negative of the class of f, and let H be a based homotopy realizing

$$0 = [f] + [g] = [fg].$$

Apply the above "bar" construction to g and H. One verifies easily that $\bar{H}$ is an $\mathcal{H}'$-concordance $\bar{f}\bar{g} \sim \mathrm{id}_{M \times \mathbb{R}^i}$. (Note: $i \geq 1$ is important here.)

b) Under the bar construction, homotopies in $\mathcal{M}(M_V, \iota)$ based at id_{M_V} determine $\mathcal{H}'$-concordances between maps in $\mathcal{H}_i(M \text{ rel } X)$, as in a). By Lemma 3.1.8, these are, therefore, $\mathcal{H}$-concordances. Thus, $f \to \bar{f}$ gives a well-defined map Φ of groups. Since $\bar{f}\bar{g} = \overline{fg}$, Φ is a homomorphism. Q.E.D.

D.6 Theorem: Φ_V is an isomorphism

$$\pi_i(\mathcal{M}(M_V, \iota), id_{M_V}) \approx \pi_i(\mathcal{H}; M \text{ rel } X).$$

We need two supporting lemmas to prove D.6. First, a definition:

D.7 Definition: We say that a map in $\mathcal{H}_i(M \text{ rel } X)$ (resp., an $\mathcal{H}$-concordance between such maps) respects parameters if it commutes with projection onto $\mathbb{R}^i$ (resp., $\mathbb{R}^i \times [0,1]$).

D.8 Lemma: a) Every $f \in \mathcal{H}_i(M \text{ rel } X)$ is $\mathcal{H}$-concordant rel X to a map that respects parameters

b) Every $\mathcal{H}$-concordance rel X between maps in $\mathcal{H}_i(M \text{ rel } X)$ that respect parameters can be replaced by one that respects parameters.

Proof: a) Use straight-line homotopies in the $\mathbb{R}^i$ factor. Adjust these to obtain stationariness near $\{0,1\}$. The result determines $\mathcal{H}'$-concordances, which, by Lemma 3.1.8, are $\mathcal{H}$-concordances.

b) Let

$$M \times \mathbb{R}^i \times [0,1] \begin{array}{l} \xrightarrow{Pr_1} M \\ \xrightarrow{Pr_2} \mathbb{R}^i \times [0,1] \end{array}$$

be the standard projections. If F is the given $\mathcal{H}$-concordance, then (Pr_1F, Pr_2) is the desired one. Q. E. D.

Now, let $h : \partial V \times [0,\infty) \to V\text{-}X$ be a homeomorphism satisfying $h(v,0) = v$ (see 3.5.1), and let

$$V_n = V - h(\partial N \times [0,n)), \qquad n \text{ a positive integer.}$$

Let $\iota_n : \partial V_n \to M_{V_n}$ be the inclusion, and define

$$j : \mathcal{M}(M_V, \iota) \to \mathcal{M}(M_{V_n}, \iota_n)$$

by letting $j(f)$ coincide with f on M_V and defining it to be the identity elsewhere.

D. 9 <u>Lemma</u>: a) <u>j is a topological imbedding onto a closed subset of $\mathcal{M}(M_{V_n}, \iota_n)$.</u>

b) <u>$j(\mathcal{M}(M_V, \iota))$ is a strong deformation retract of $\mathcal{M}(M_{V_n}, \iota_n)$.</u>

<u>Proof</u>: Statement a) is an elementary exercise in general topology.

To prove b), let

$$g : \partial V_n \times [0,1] \to M_{V_n}$$

be a topological collar satisfying $g(x,0) = x$ and $g(\partial V_n \times 1/2) = \partial V$. Let $M_s = M\text{-}\mathrm{int}\, g(\partial V_n \times [0,s])$, so that $M_0 = M_{V_n}$ and $M_{1/2} = M_V$.

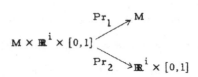

For $0 \leq t \leq 1$, define the homeomorphism

$$H_t : M_{t/2} \to M_0$$

by

$$H_t(g(x,s)) = g(x, \frac{2s-t}{2-t}), \quad s \geq 1/2t$$

$$H_t(y) = y \quad , \quad y \notin \text{image } g.$$

Then, for any $f \in \mathcal{M}(M_{V_n}, \iota_n)$, define

$$K_t(f)(y) = \begin{cases} H_t^{-1} f H_t(y), & y \in M_{t/2} \\ y & , \quad y \notin M_{t/2} \end{cases}.$$

Note that $2s-t/2-t \leq s$, so that if $f | (M_{V_n} - \text{int } M_V)$ is the identity, the same is true for $K_t(f) | M_{V_n} - \text{int } M_V$, for all $0 \leq t \leq 1$. Therefore,

$$K_t : (\mathcal{M}(M_{V_n}, \iota_n), \text{image } j) \to (\mathcal{M}(M_{V_n}, \iota_n), \text{image } j)$$

is a deformation of pairs, and

$$K_1(\mathcal{M}(M_{V_n}, \iota_n)) \subseteq \text{image } j.$$

Since $\mathcal{M}(M_{V_n}, \iota_n)$ is an ANR and image j is a closed subset, a theorem of Fox [8] implies the desired result. Q. E. D.

Proof of Theorem D. 6:

To prove that Φ_V is onto, choose any $f \in \mathcal{H}_i(M \text{ rel } X)$. We may assume, by D. 8 that it respects parameters, and we may clearly suppose that

$$\mathrm{supp}\, f \subseteq (M-X) \times D^i,$$

where $D^i \subseteq \mathbb{R}^i$ is the unit disc. It follows that

$$\mathrm{supp}\, f \subseteq M_{V_n} \times D^i,$$

for n sufficiently large, so that f represents a class in

$$\Phi_{V_n}(\pi_i(\mathcal{M}(M_{V_n}, \iota_n)), \mathrm{id}_{M_{V_n}}).$$

Since $\Phi_V = \Phi_{V_n} \circ j_*$, where j_* is induced by j above, and since j_* is an isomorphism (D. 9), f represents a class in image Φ_V.

Similarly, if there is an $\mathcal{H}$-concordance between maps coming from $\mathcal{M}(M_V, \iota)$, we can show--using compactness of support--that these maps determine homotopic classes in $\pi_i(\mathcal{M}(M_{V_n}, \iota_n), \mathrm{id}_{M_{V_n}})$, n suitably large, and then apply D. 9 (in its strong form) to conclude that they determine homotopic classes in $\pi_i(\mathcal{M}(M_V, \iota), \mathrm{id}_{M_V})$. Q. E. D.

D. 10 Proposition: *Let M, X, and V be as above, and choose any $i \geq 1$. Suppose that $\partial(M_V \times D^i)$ is connected. Then*

$$\pi_i(\mathcal{H}; M \text{ rel } X) = 0$$

provided that a sequence of obstructions

$$o_j \in \tilde{H}^j(\Sigma^i(M/X); \{\pi_j(M-X)\}), \quad j \geq 2,$$

vanish.

Here, $\{\pi_j(M-X)\}$ is the natural local-coefficient system. The proof is exactly the same as that of Proposition 3.4.3, which this corollary generalizes. The condition on $\partial(M_V \times D^i)$ implies that the pointed set

$$\pi_1(M_V \times D^i, \partial(M_V \times D^i))$$

is trivial, and this allows us to avoid the problems with o_1 that arise when $\pi_1(M_V \times D^i)$ is non-abelian.

In general, the homotopy groups of

$$\pi_i(\mathcal{M}(M_V, \iota), \mathrm{id}_{M_V}), \quad i \geq 1 \tag{1}$$

are difficult to compute. In contrast, the groups

$$\pi_i(\mathcal{M}(M_V, *), *),$$

where $* : \partial V \to M_V$ is any constant map, as well as its constant extension to M_V, are easily identified with

$$[\Sigma^i(M/X), M-X]. \tag{2}$$

These latter groups can often be given explicitly (see the examples in D.17). We now prove a theorem which gives conditions under which the groups (1) are isomorphic to those in (2) above.

D.11 <u>Theorem</u>: <u>Suppose that</u> $\mathrm{Map}(M_V, M_V)$ <u>and</u> $\mathrm{Map}(\partial V, M_V)$ <u>admit the structure of H-spaces (with homotopy-associative multiplication, strict-identities, and homotopy inverses), and suppose that the fibre</u>

map $p : \mathrm{Map}(M_V, M_V) \to \mathrm{Map}(\partial V, M_V)$ *respects the multiplications. Then, there is a basepoint-preserving homotopy equivalence*

$$(\mathcal{M}(M_V, *), *) \to (\mathcal{M}(M_V, \iota), \mathrm{id}_{M_V}),$$

where $*$ *is a constant map, as described above.*

Proof: Choose any $g \in \mathrm{Map}(M_V, M_V)$. We then get a commutative diagram

$$\begin{array}{ccc} \mathrm{Map}(M_V, M_V) & \xrightarrow{g^{\#}} & \mathrm{Map}(M_V, M_V) \\ p \downarrow & & \downarrow p \\ \mathrm{Map}(\partial V, M_V) & \xrightarrow{p(g)^{\#}} & \mathrm{Map}(\partial V, M_V), \end{array}$$

where $g^{\#}(f) = g \cdot f$ and $p(g)^{\#}(h) = p(g) \cdot h$, $\cdot$ being the H-space operation. Because the operation has homotopy-inverses, $g^{\#}$ and $p(g)^{\#}$ are homotopy equivalences, and so they induce isomorphisms of homotopy groups. Therefore, applying the Five Lemma to the exact homotopy sequence of the fibration, together with Whitehead's theorem and the fact that fibres are ANR's, we conclude that

$$g^{\#} : (p^{-1}(p(f)), f) \to (p^{-1}(p(g \cdot f)), g \cdot f)$$

is a based homotopy-equivalence. Let e be the (strict) identity of $\mathrm{Map}(M_V, M_V)$; then $p(e)$ is a strict identity for $\mathrm{Map}(\partial N, M_V)$. Setting $f = e$, we get

$$g^{\#} : (\mathcal{M}(M_V, p(e)), e) \to (\mathcal{M}(M_V, p(g)), g)$$

is a based homotopy-equivalence. The desired homotopy-equivalence is, then, the composite

$$(\mathrm{id}_{M_V})^{\#} \circ [(*)^{\#}]^{-1}.$$ Q. E. D.

D.12 Corollary (Proposition 3.4.2): Suppose that M-X admits the structure of a homotopy-associative H-space with homotopy-identity and homotopy-inverses. Then, for $i \geq 1$,

$$\pi_i(\mathcal{H}; M \text{ rel } X) \approx (\Sigma^i(M/X),\ M\text{-}X].$$

Proof: We endow M_V with an H-space structure via the inclusion $M_V \subsetneq M\text{-}X$, which is a homotopy equivalence. An easy application of the homotopy extension property allows us to modify the multiplication so that the identity becomes a strict identity. Then, use the operation on M_V to give $\mathrm{Map}(M_V, M_V)$ and $\mathrm{Map}(\partial V, M_V)$ H-space structures. The result now follows from D.11 and the preceding remarks. Q. E. D.

D.13 Corollary: Suppose that M-X has the homotopy type of a wedge of spheres. Then, for $i \geq 1$,

$$\pi_i(\mathcal{H}; M \text{ rel } X) \approx [\Sigma^i(M/X),\ M\text{-}X].$$

Proof: If M-X has the homotopy type of a wedge of spheres, then so does M_V. The co-H-space structure of such a wedge induces an H-space structure on $\mathrm{Map}(M_V, M_V)$ and one on $\mathrm{Map}(\partial V, M_V)$. The operations are compatible with p and are homotopy-associative; there are homotopy-identities and homotopy-inverses. It is an interesting exercise involving the homotopy extension and covering homotopy extension properties to show that the operations may be modified, compatibly with p, so as to make the identities strict. We leave this to the reader.

Q. E. D.

D.14 Corollary: If M-X is contractible, then

$$\pi_i(\mathcal{H}; M \text{ rel } X) = 0, \quad i \geq 1.$$

This follows immediately from D.12.

D.15 Corollary: if M has the homotopy type of a topological group, then

$$\pi_i(\mathcal{H}; M) \approx [\Sigma^i(M_+), M], \quad i \geq 1,$$

where $M_+ = M \cup \text{pt}$.

For example, when M is a homotopy n-torus,

$$\pi_i(\mathcal{H}; M) = \begin{cases} 0 & , \quad i > 1 \\ \pi_1(M), & i = 1, \end{cases}$$

sharpening the abelian case of Proposition 3.4.3.

D.16 Corollary: Let $X \to M \xrightarrow{\pi} S^n$ be a fibre bundle with X admitting a homotopy-associative H-space structure with homotopy identity and homotopy inverses. Then, for $i \geq 1$,

$$\pi_i(\mathcal{H};M \text{ rel } X) \approx [\Sigma^{i+n}X, X] + \pi_{i+n}(X).$$

Proof: M-X is homeomorphic to $X \times \mathbb{R}^n$ which inherits the H-space structure of X (or of the product). Thus, by D.12,

$$\pi_i(\mathcal{H};M \text{ rel } X) \approx [\Sigma^i(M/X), X \times \mathbb{R}^n]$$

$$\approx [\Sigma^i(M-\pi^{-1}(\text{int } D^n_+)/\pi^{-1}(\partial D^n_+), X]$$

$$\approx [\Sigma^i(X \times D^n / X \times \partial D^n), X]$$

$$\approx [\Sigma^i(\Sigma^n(X_+)), X], \qquad \text{Q.E.D.}$$

D.17 Examples:

a) Let M be an (n-1)-connected 2n-manifold. Then, $M-D^{2n}$ has the homotopy type of a wedge of, say, r spheres of dimension n. Then, for $i \geq 1$,

$$\pi_i(\mathcal{H};M \text{ rel } D^{2n}) \approx [\Sigma^i M, S^n_1 \vee \dots \vee S^n_r].$$

Note that, in general, this differs from the direct sum of r copies of $[\Sigma^i M, S^n]$.

b) $\pi_i(\mathcal{H};S^p \times S^q \text{ rel } D^{p+q}) \approx \pi_{p+q+i}(S^p \vee S^q) + \pi_{p+i}(S^p \vee S^q) + \pi_{q+i}(S^p \vee S^q)$, for $i \geq 1$. This example is similar to that in a). Note that $\Sigma^i(S^p \times S^q) \sim S^{p+q+i} \vee S^{p+i} \vee S^{q+i}$ for $i \geq 1$.

c) Any principal fibre bundle $\pi : M \to S^n$ with fibre X satisfies the hypotheses of D.16. Thus:

$$\pi_i(\mathcal{H};SO_{n+1}\text{ rel }SO_n) \approx [\Sigma^{i+n}SO_n, SO_n] + \pi_{i+n}(SO_n)$$

and $$\pi_i(\mathcal{H};U_{k+1}\text{ rel }U_k) \approx [\Sigma^{i+2k+1}U_k, U_k] + \pi_{i+2k+1}(U_k).$$

d) M is an S^3 bundle over S^n:

$$\pi_i(\mathcal{H};M\text{ rel }S^3) \approx \pi_{i+n+3}(S^3) + \pi_{i+n}(S^3).$$

e) M is the n-torus T^n, and $\pi : T^n \to S^1$ is the standard projection onto the last factor:

$$\pi_i(\mathcal{H};T^n\text{ rel }T^{n-1}) = 0, \quad i \geq 1.$$

f) M is a homotopy n-sphere:

$$\pi_i(\mathcal{H};M) \approx \pi_{i+n}(S^n) + \pi_i(S^n), \quad i \geq 1.$$

APPENDIX E: Some Applications to the Groups $\pi_i(\mathrm{Diff}(M^n, X))$

In this appendix, we apply the machinery developed in the text to sharpen and expand some of the results of Chapter 2, [I]. Here, M^n is a closed, oriented C^∞ n-manifold, and X is a compact, C^∞ submanifold of codimension-0. Recall that $\mathrm{Diff}(M^n, X)$ consists of all orientation-preserving diffeomorphisms $M^n \to M^n$ that fix each point of X.

E.1 The main diagram of Chapter 2

In Chapter 2, we construct a commutative diagram:

$$\begin{array}{ccc} \pi_i(\mathrm{Diff}(S^n, D^n_+)) & \xrightarrow{E_*} & \pi_i(\mathrm{Diff}(M^n, X)) \\ \Phi\downarrow & & \downarrow\Phi \\ \pi_i(\mathcal{D}iff; S^n \mathrm{rel}\ D^n_+) & \xrightarrow{\mathcal{E}_*} & \pi_i(\mathcal{D}iff; M^n \mathrm{rel}\ X)\ . \end{array} \tag{1}$$

(Actually, in Chapter 2, we only consider the case $X = \phi$ for the diagram, but the definitions and verification of commutativity are the same.) Here, the homomorphisms Φ are "forgetful" homomorphisms, and the homomorphism $\mathcal{E}_*$ is just the homomorphism induced by the relative $\mathcal{D}iff$-map

$$(M^n, X) \to (S^n, D^n_+)$$

described in 3.3.6.

Recall that there is an isomorphism

$$\pi_i(\mathcal{D}iff; S^n \ \mathrm{rel}\ D^n_+) \approx \Gamma^{n+i+1}$$

that takes $\Phi(\pi_i(\mathrm{Diff}(S^n, D^n_+)))$ onto the Gromoll subgroup Γ^{n+i+1}_{i+1} (see [1], Chapter 1). In [1], Chapter 1, we obtain certain non-triviality results for Γ^{n+i+1}_{i+1} that we use in Chapter 2 and shall use here.

In particular, we are interested in showing that image E_* is non-trivial. By (1), it is sufficient to show that

$$(2) \qquad \Gamma^{n+i+1}_{i+1} \not\subset \ker \mathcal{E}_* .$$

E. 2 The augmented diagram

We enlarge diagram (1) as follows:

$$(3) \qquad \begin{array}{ccc} \pi_i(\mathrm{Diff}(S^n, D^n_+)) & \xrightarrow{E_*} & \pi_i(\mathrm{Diff}(M^n, X)) \\ \Phi\downarrow & & \downarrow\Phi \\ \pi_i(\mathcal{Diff}; S^n \mathrm{rel}\ D^n_+) & \xrightarrow{\mathcal{E}_*} & \pi_i(\mathcal{Diff}; M\ \mathrm{rel}\ X) \\ \partial\uparrow\approx & & \uparrow\partial' \\ \pi_{i+1}(\mathcal{B}, \mathcal{Diff}; S^n \mathrm{rel}\ D^n_+) & \xrightarrow{\mathcal{E}'_*} & \pi_{i+1}(\mathcal{B}, \mathcal{Diff}; M^n \mathrm{rel}\ X) \\ & & \uparrow k \\ & & \pi_{i+1}(\mathcal{B}; M^n \mathrm{rel}\ X). \end{array}$$

Here, $\mathcal{B}$ is any geometric category other than $\mathcal{Diff}$, and ∂' and k (resp., ∂) are maps in the $(\mathcal{B}, \mathcal{Diff})$-sequence for $M^n \mathrm{rel}\ X$ (resp., $S^n \mathrm{rel}\ D^n_+$). The commutativity of the bottom square follows from the Second Naturality Theorem. That ∂ is an isomorphism follows from the Exactness Theorem, together with the fact (3.4.1) that $\pi_i(\mathcal{B}; S^n \mathrm{rel}\ D^n_+) = 0$, for all $i \geq 0$.

E.3 Smooth homotopy-tori

We specialize $\mathcal{B}$, M^n, and X in diagram (3) as follows: M^n is a $\mathcal{Diff}$-manifold having the homotopy type of the n-torus T^n; $X = \phi$; $\mathcal{B} = \mathcal{PL}$.

It is proved in [16], that M^n is parallelizable. Therefore, we may apply 3.5.13 a) to conclude that

(4) $$\mathcal{E}'_* \text{ is a split-injection}, \quad n+i \geq 6 .$$

(Alternatively, without [16], we could use 3.5.12.) Moreover, Corollary 3.5.9 states:

(5) $$\pi_{i+1}(\mathcal{PL}; M^n) = \begin{cases} \mathbb{Z}_2, & i = 1 \\ 0, & i \geq 2 \end{cases} \quad n+i \geq 5$$

Thus, we conclude that, for $n+i \geq 6$,

(6) $$\partial' \begin{cases} \text{is an isomorphism, when } i > 2 \\ \text{is an injection with } 0 \text{ or } \mathbb{Z}_2 \text{ cokernel, when } i = 2 \\ \text{has } \mathbb{Z}_2 \text{ or } 0 \text{ kernel, when } i = 1 . \end{cases}$$

Collecting (4) - (6), we conclude:

E.4 Proposition: Let M^n be a smooth, homotopy n-torus. Then, the homomorphism

$$\Gamma^{n+i+1} \approx \pi_i(\mathcal{Diff}; S^n \text{rel } D^n_+) \xrightarrow{\mathcal{E}_*} \pi_i(\mathcal{Diff}; M^n)$$

a) is an injection, when $i \geq 2$, $n+i \geq 5$

b) has kernel 0 or $\mathbb{Z}_2$, when $i = 1$, $n+i \geq 5$.

Referring to (2) and the comments that precede it, we see that E_* is non-trivial provided that

a') Γ^{n+i+1}_{i+1} is non-trivial, when $i \geq 2$

b') Γ^{n+i+1}_{i+1} has more than two elements, when $i = 1$.

More precisely, we have:

E.5 Corollary: Let M^n be a smooth, homotopy n-torus, and let E_* be as in diagram (2). When $i \geq 2$, $n+i \geq 5$, image E_* is an extension of Γ^{n+i+1}_{i+1}. When $i = 1$, $n+1 \geq 5$, image E_* is an extension of $\Gamma^{n+2}_2 \approx \Gamma^{n+2}$ or of $\Gamma^{n+2}/\mathbb{Z}_2$.

Remark: Since E_* factors through $\pi_i(\mathrm{Diff}(M^n, X))$, for any $X \neq M^n$, we obtain the same result when M^n is replaced by (M^n, X), $X \neq M^n$.

We now want to obtain some additional information on kernel $\mathcal{E}_*$. Of course, this is of interest only when $i = 1$.

E.6 Lemma: kernel $\mathcal{E}_* \subseteq bP_{n+i+2}$

Proof: Since ∂ is an isomorphism, it suffices to show that $\partial(\ker(\partial' \circ \mathcal{E}'_*)) = bP_{n+i+2}$, or that $\partial(\mathcal{E}'^{-1}_*(\text{image } k)) \subseteq bP_{n+i+2}$. Now, in fact, the identifications of Γ^{n+i+1} with $\pi_i(\mathcal{D}iff\, ;S^n \text{rel } D^n_+)$ and with

$\pi_{i+1}(\mathcal{B}, \mathcal{D}iff; S^n \text{rel } D^n_+)$ are compatible with ∂, so that it suffices to show that

$$\mathcal{E}'^{-1}_*(\text{image } k) \subseteq bP_{n+i+2}.$$

Consider the commutative diagram, $i \geq 1$,

$$\begin{array}{ccccc}
 & & \pi_{i+1}(\mathcal{PL}, \mathcal{D}iff; S^n, D^n_+) & \xrightarrow{\approx} & \pi_{i+1}(\mathcal{H}, \mathcal{D}iff; S^n, D^n_+) \\
 & & \downarrow \mathcal{E}'_* & & \downarrow \mathcal{E}''_* \\
\pi_{i+2}(\mathcal{H}, \mathcal{PL}; T^n) & \rightarrow & \pi_{i+1}(\mathcal{PL}, \mathcal{D}iff; T^n) & \xrightarrow{\ell} & \pi_{i+1}(\mathcal{H}, \mathcal{D}iff; T^n) \\
 & \searrow^{\approx} & \uparrow k & & \\
 & & \pi_{i+1}(\mathcal{PL}, T^n). & &
\end{array}$$

The horizontal isomorphism (which again preserves bP_{n+i+2}) is easy to deduce from 3.4.1. The isomorphism

$$\pi_{i+2}(\mathcal{H}, \mathcal{PL}; T^n) \xrightarrow{\approx} \pi_{i+1}(\mathcal{PL}, T^n), \quad i \geq 1,$$

follows from the fact (3.4.3) that $\pi_{i+1}(\mathcal{H}; T^n) = 0$, $i \geq 1$. Therefore,

$$\begin{aligned}
\mathcal{E}'^{-1}_*(\text{image } k) &= \mathcal{E}'^{-1}_*(\text{kernel } \ell) \\
&= \text{kernel } (\ell, \mathcal{E}'_*) \\
&= \text{kernel } \mathcal{E}''_* \subseteq bP_{n+i+2}.
\end{aligned}$$

The last inclusion is given by 3.5.13 b). Q.E.D.

We now refer the reader to [1], Chapter 1, Theorem 1.4.4 and Appendix A.1. These imply that

a) <u>For every</u> $n \geq 9$, <u>there exists an</u> $i \geq 2$ <u>such that</u>

$$\Phi : \pi_i(\text{Diff}(S^n, D^n_+)) \rightarrow \Gamma^{n+i+1}_{i+1} \subseteq \Gamma^{n+i+1}$$

is non-trivial (Theorem 1.4.4).

b) *For* $n = 7, 8$, $\pi_1(\mathrm{Diff}(S^n, D^n_+))$ *has torsion elements that do not belong to* $\Phi^{-1}(bP_{n+3})$ *(Appendix A.1).*

We may, therefore, conclude that:

a') *For every* $n \geq 9$, *there exists an* $i \geq 2$ *such that*

$$E_* : \pi_i(\mathrm{Diff}(S^n, D^n_+)) \to \pi_i(\mathrm{Diff}\ M^n)$$

is non-trivial.

b') *For* $n = 7, 8$

$$\underline{\text{image}}\,(E_* : \pi_1(\mathrm{Diff}(S^n, D^n_+)) \to \pi_1(\mathrm{Diff}\ M^n))$$

has non-trivial torsion.

E.7 *Corollary: Let* M^n *be a smooth, homotopy n-torus. For every* $n \geq 7$, M^n *and* $\mathrm{Diff}_0 M^n$ *(the identity component of* $\mathrm{Diff}\ M^n$*) have distinct homotopy types.*

This follows immediately from the facts that $\pi_i(M^n)$ is free-abelian for all i and 0 for $i \geq 2$.

For $n = 1, 2$, E.7 is false (cf. [1], Introduction). When $3 \leq n \leq 6$, the situation is undecided.

E.8 *Corollary: Suppose that* $n = 4k$, $k \geq 3$, *or* $n = 8k-6$, k *not a power of two. Then,* $\mathrm{Diff}_0 M^n$ *does not have the homotopy type of a finite CW complex.*

Proof: We use Theorems 1.4.3 and 1.4.4 of [1], together with E.5 above, applied to the case $i = 2$. Thus, image E_* is an extension of Γ_3^{n+3}, which, in the above-stated dimensions, is non-zero (1.4.3, 1.4.4 of [1]). Therefore, $\pi_2(\mathrm{Diff}\, M^n) \neq 0$ in these dimensions, so that, by a result of Browder [3], $\mathrm{Diff}_0 M^n$ cannot have finite type. Q.E.D.

Those two corollaries comprise Theorem C of [1].

Let

$$T_+ = \underbrace{S^1 \times \ldots \times S^1}_{n-1} \times D^1_+ \subset \underbrace{S^1 \times \ldots \times S^1}_{n} = T^n.$$

We prove in [1], Chapter 2, that $\mathrm{Diff}(T^n, T_+)$ is homotopy-abelian. From this and other results in that chapter, we conclude (Corollary 2.6.2) that

$$\mathrm{Diff}_0(T^n, T_+)$$

does not have finite type, provided $n \geq 25$. We can now improve this dimension restriction.

E.9 Corollary: If $n \geq 7$, then $\mathrm{Diff}_0(T^n, T_+)$ does not have finite type.

Proof: If $\mathrm{Diff}_0(T^n, T_+)$ does have finite type, then it must be contractible or have the homotopy type of a torus [16a]. In either case, $\pi_i(\mathrm{Diff}_0(T^n, T_+)) = 0$, $i \geq 2$, and $\pi_1(\mathrm{Diff}_0(T^n, T_+))$ has no torsion. This contradicts E.5 (applied to (T^n, T_+)--see the remark following E.5) together with conclusions a') and b') preceding E.7. Q.E.D.

Remark: $\mathrm{Diff}(T^n, T_+)$ may be identified with the group of C^∞ pseudo-isotopies between $\mathrm{id}_{T^{n-1}}$ and itself.

E.10 Manifolds that are $K(\pi, 1)$'s

According to 3.4.3, if M^n is a $K(\pi, 1)$

$$\pi_i(\mathcal{H}; M^n) = 0, \quad i > 1,$$

so that the boundary homomorphism

$$\pi_{i+1}(\mathcal{H}, \mathcal{D}iff; M^n) \xrightarrow{\partial'} \pi_i(\mathcal{D}iff; M^n)$$

is an isomorphism for $i > 1$ and an injection for $i = 1$. Suppose now that the tangent bundle of M^n is stably fibre-homotopy trivial (e.g., parallelizable). Then, we may apply 3.5.13 b), which states that

$$\mathrm{kernel}(\pi_{i+1}(\mathcal{H}, \mathcal{D}iff; S^n \mathrm{rel}\ D^n_+) \xrightarrow{\mathcal{E}'_*} \pi_{i+1}(\mathcal{H}, \mathcal{D}iff; M^n))$$

is contained in bP_{n+i+2}.

Thus, we obtain:

E.10 A non-triviality criterion: Let M^n be as above. Then,

$$\pi_i(\mathrm{Diff}(S^n, D^n_+)) \xrightarrow{E_*} \pi_i(\mathrm{Diff}\ M^n)$$

is non-trivial whenever $\Gamma^{n+i+1}_{i+1} \not\subset bP_{n+i+2}$.

This follows immediately from the remarks above and diagram (3). In [1], §1.3, we detect non-trivial Γ^k_{2Q-2}, for certain Q and k, using the τ-pairing of Munkres. In fact, our detection procedure shows that the composition of inclusion and projection

$$\Gamma^k_{2Q-2} \subset \Gamma^k \to \Gamma^k / bP_{k+1},$$

contains a copy of the cyclic group $\mathbb{Z}_Q$ in its image.

Now it follows from the remarks preceding E.10 that

$$\mathcal{E}_*(\pi_i(\mathcal{D}iff; S^n \text{ rel } D^n_+))$$

is an extension of $\Gamma^{n+i+1}/bP_{n+i+2}$. Therefore, we may conclude that, for appropriate i, n, Q, image E_* contains a copy of $\mathbb{Z}_Q$. We now make this more precise.

Let i, n, and Q be any integers satisfying:

(i) Q is an odd prime

(ii) $i \leq 2Q-3$

(iii) $n+i = 2(uQ+v+1)(Q-1) - 2(u-v) - 2$,

for some u, v such that $0 \leq v < u \leq Q-1$, $u-v \neq Q-1$.

E.11 Corollary: Let i, n, Q be as above, and let M^n be a $K(\pi, 1)$ with stably fibre-homotopy trivial tangent bundle. Then,

$$\pi_i(\text{Diff } M^n)$$

contains a copy of $\mathbb{Z}_Q$.

In particular, if we set $i = 2$ and let n assume the corresponding

$1/2\, Q(Q-1)-1$ values allowed by (iii) above, we conclude again that for such M^n, $\mathrm{Diff}_0 M^n$ does not have finite type.

The restraints (i) - (iii) on i, n, and Q derive from conditions in results of Toda (see [1], 1. 3. 4) on the stable homotopy groups of spheres.

Notice that the method of Chapter 2, [1], uses the fact that

$$\text{kernel } \mathcal{E}_* \cap bP_{n+i+2} \subseteq I(M^n \times S^{i+1}) \cap bP_{n+i+2}$$

and then proceeds by showing that, under certain conditions

(i) $I(M^n \times S^{i+1}) \cap bP_{n+i+2}$ is small,

(ii) $\Gamma^{n+i+1}_{i+1} \cap bP_{n+i+2}$ is large.

It follows that not all of Γ^{n+i+1}_{i+1} is contained in kernel $\mathcal{E}_*$, as desired.

Our method here begins with the fact that, under certain conditions (overlapping, but not containing or contained in, the above),

$$\text{kernel } \mathcal{E}_* \subseteq bP_{n+i+2}$$

and then discovers certain n and i for which

$$\Gamma^{n+i+1}_{i+1} / \left(\Gamma^{n+i+1}_{i+1} \cap bP_{n+i+2}\right)$$

is non-trivial.

In some sense, then, the methods are complementary.

The conditions of applicability for both methods are satisfied by flat manifolds and by oriented manifolds of constant negative curvature whose tangent bundles are stably fibre-homotopy trivial.

REFERENCES

[1] P. Antonelli, D. Burghelea, P. J. Kahn, The non-finite homotopy type of some diffeomorphism groups, Topology, to appear.

[2] J. Boardman, R. Vaught, Homotopy-everything H-spaces, Bull. AMS, 74(1968), 1117-1123.

[3] W. Browder, Torsion in H-spaces, Ann. of Math., 74(1961), 27-51.

[4] W. Browder, Surgery on simply connected manifolds (mimeographed notes), Princeton Univ., 1969.

[5] M. Brown, Locally flat imbeddings of topological manifolds, Ann. of Math., 75(1962), 331-341.

[6] J. Cerf, Topologie de certaines espaces de plongements, Bull. Soc. Math. de France, 89(1961), 227-380.

[7] A. Douady, Seminaire H. Cartan 1961/1962, No. 1.

[8] R. Fox, On homotopy type and deformation retracts, Ann. of Math. 44(1943), 40-50.

[9] A. Haefliger, C. T. C. Wall, Piecewise linear bundles in the stable range, Topology 4(1965), 209-214.

[10] M. Hirsch, On normal microbundles, Topology, 5(1966), 229-240.

[11] M. Hirsch, "Smoothings of Piecewise Linear Manifolds, I" (mimeographed), U. of Cal.

[12] M. Hirsch, B. Mazur, "Smoothings of Piecewise Linear Manifolds" (mimeographed), Cambridge U., 1964.

[13] M. Hirsch, On combinatorial submanifolds of differentiable manifolds, Comment. Math. Helv. 36(1961), 103-111.

[14] J. Hodgson, Automorphisms of thickenings, Bull. AMS 73(1967), 678-681.

[15] J. Hodgson, Poincaré complex thickenings, Bull. AMS 76(1970), 1039-1043.

[16] W. C. Hsiang, J. Shaneson Fake Tori, "Topology of Manifolds", Markham Press, Chicago, 1971.

[16a] J. Hubbuck, On homotopy commutative H-spaces, Topology 8(1969), 119-126.

[16b] J. Hudson, "Piecewise Linear Topology," Benjamin Press, New York, 1969.

[17] P. J. Kahn, Triangulation and smoothing, Chs. 1 and 2, (mimeographed), Inst. for Adv. Study, Princeton, 1969.

[18] R. Kirby, L. Siebenmann, On the triangulation of manifolds and the Hauptvermutung, Bull. AMS, 75(1969), 742-750.

[19a] R. Kirby, "Lectures on triangulation of manifolds", (mimeographed notes), UCLA, 1969.

[19b] R. Kirby, Stable homeomorphisms and the annulus conjecture, preprint.

[20] R. Kirby, L. Siebenmann, For manifolds the Hauptvermutung and the triangulation conjecture are false, AMS Notices, 1969.

[21] J. Kister, Microbundles are fibre bundles, Ann. of Math., 80(1964), 190-199.

[22] N. Kuiper, R. Lashof, Microbundles and bundles, I, Invent. Math., 1(1966), 1-17.

[23] C. Kuratowski, Sur les espaces localement connexes et péaniens en dimension n, Fund. Math. 24(1935), 269-287.

[24] R. Lashof, M. Rothenberg, Microbundles and smoothing, Topology 3(1965), 357-388.

[25] R. Lashof, M. Rothenberg, Hauptvermutung for manifolds, "Topology of manifolds", Prindle, Weber, and Schmidt, 1968, Boston,

[26] J. Milnor, Differentiable manifolds which are homotopy spheres (mimeographed), Princeton U., 1956.

[27] J. Milnor, Microbundles and differentiable structures [PL], (mimeographed), Princeton U., 1961.

Microbundles, I [topological], Topology 3(1964), 53-80.

[28] C. Morlet, Voisinages tubulaires des variétés semi-lineaires, C.R. Acad. Sci. Paris 262(1966), 740-743.

[29] J. Munkres, "Elementary Differential Topology", Chapter 2, Annals of Math. Studies 54, Princeton U. Press.

[30] L. Siebenmann, Disruption of low-dimensional handlebody theory by Rohlin's theorem, "Topology of manifolds", Markham Press, 1969, Chicago.

[31] E. Spanier, "Algebraic Topology", McGraw-Hill, 1966, New York.

[32] M. Spivak, Spaces satisfying Poincaré duality, Topology, 6(1967), 77-102.

[33] D. Sullivan, Thesis, Princeton University, 1965.

[34] D. Sullivan, a) Triangulating homotopy equivalences

b) Smoothing homotopy equivalences,

(mimeographed), U. of Warwick, 1966.

[35] D. Sullivan, Geometric topology seminar notes (mimeographed), Princeton Univ., 1967.

[36] J. Wagonner, Thesis, Princeton University, 1965.

[37] C. T. C. Wall, Surgery on compact manifolds, (mimeographed notes) University of Liverpool.

[38] C. T. C. Wall, On homotopy tori and the annulus theorem, Bull. London Math. Soc. 1(1969), 95-97.

[39] J. H. C. Whitehead, On C^1-complexes, Annals of Math. 41(1940), 809-824.